CODING AND MODULATION FOR DIGITAL TELEVISION

MULTIMEDIA SYSTEMS AND APPLICATIONS SERIES

Consulting Editor

Borko Furht
Florida Atlantic University

Recently Published Titles:

CELLULAR AUTOMATA TRANSFORMS: *Theory and Applications in Multimedia Compression, Encryption, and Modeling*, by Olu Lafe
        ISBN: 0-7923-7857-1

COMPUTED SYNCHRONIZATION FOR MULTIMEDIA APPLICATIONS, by Charles B. Owen and Fillia Makedon
        ISBN: 0-7923-8565-9

STILL IMAGE COMPRESSION ON PARALLEL COMPUTER ARCHITECTURES, by Savitri Bevinakoppa
        ISBN: 0-7923-8322-2

INTERACTIVE VIDEO-ON-DEMAND SYSTEMS: *Resource Management and Scheduling Strategies*, by T. P. Jimmy To and Babak Hamidzadeh
        ISBN: 0-7923-8320-6

MULTIMEDIA TECHNOLOGIES AND APPLICATIONS FOR THE 21st CENTURY: *Visions of World Experts*, by Borko Furht
        ISBN: 0-7923-8074-6

INTELLIGENT IMAGE DATABASES: *Towards Advanced Image Retrieval,* by Yihong Gong
        ISBN: 0-7923-8015-0

BUFFERING TECHNIQUES FOR DELIVERY OF COMPRESSED VIDEO IN VIDEO-ON-DEMAND SYSTEMS, by Wu-chi Feng
        ISBN: 0-7923-9998-6

HUMAN FACE RECOGNITION USING THIRD-ORDER SYNTHETIC NEURAL NETWORKS, by Okechukwu A. Uwechue, and Abhijit S. Pandya
        ISBN: 0-7923-9957-9

MULTIMEDIA INFORMATION SYSTEMS, by Marios C. Angelides and Schahram Dustdar
        ISBN: 0-7923-9915-3

MOTION ESTIMATION ALGORITHMS FOR VIDEO COMPRESSION, by Borko Furht, Joshua Greenberg and Raymond Westwater
        ISBN: 0-7923-9793-2

VIDEO DATA COMPRESSION FOR MULTIMEDIA COMPUTING, edited by Hua Harry Li, Shan Sun, Haluk Derin
        ISBN: 0-7923-9790-8

REAL-TIME VIDEO COMPRESSION: *Techniques and Algorithms,* by Raymond Westwater and Borko Furht
        ISBN: 0-7923-9787-8

MULTIMEDIA DATABASE MANAGEMENT SYSTEMS, by B. Prabhakaran
        ISBN: 0-7923-9784-3

CODING AND MODULATION FOR DIGITAL TELEVISION

by

Gordon Drury

Garik Markarian

Keith Pickavance

KLUWER ACADEMIC PUBLISHERS
Boston / Dordrecht / London

Distributors for North, Central and South America:
Kluwer Academic Publishers
101 Philip Drive
Assinippi Park
Norwell, Massachusetts 02061 USA
Telephone (781) 871-6600Fax (781) 681-9045
E-Mail <kluwer@wkap.com>

Distributors for all other countries:
Kluwer Academic Publishers Group
Distribution Centre
Post Office Box 322
3300 AH Dordrecht, THE NETHERLANDS
Telephone 31 78 6392 392   Fax 31 78 6392 254
E-Mail <services@wkap.nl>

  Electronic Services <http://www.wkap.nl>
ISBN 978-1-4419-5005-5   e-ISBN 978-0-306-47036-3

Library of Congress Cataloging-in-Publication Data

Drury, Gordon.
    Coding and modulation for digital television / by Gordon Drury, Garik Markarian,
Keith Pickavance.
        p. cm. -- (Multimedia systems and applications series)
    Includes bibliographical references and index.

    1. Digital television. 2. Coding theory. 3. Digital modulation. 4. Multiplexing I.
Markarian, Garik. II. Pickavance, Keith, 1970- III. Title. IV. Multimedia systems
and applications.

TK6678.D78 2000
621.388--dc21   00-062479

Printed on acid-free paper.   Printed in the United States of America

Ideal for communication systems, AHA's Astro OC-3 provides up to 3 dB of coding gain and
is the only commercially available turbo code technology to achieve channel rates of 200
Mbit/sec. This front cover photograph was printed with permission from Advanced Hardware
Architectures.

*The Publisher offers discounts on this book for course use and bulk purchases.
For further information, send email to <scott.delman@wkap.com>.*

Contents

Preface

For over 50 years analogue TV systems, such as NTSC, PAL and SECAM, have proved to be very rugged and reliable, providing reasonable quality pictures and audio even in severe environments. However, the world's consumers paid a high price for the diversity of analogue broadcasting standards. The division of the world television services between NTSC, PAL and SECAM built substantial barriers not only between the different continents, but also between countries of the same continent.

The latest advances of technology attempt to remove this diversity by providing the transmission of digital television (DTV) signals in the same bandwidth, currently used by the analogue signal. Such a transmission can offer improved quality video and audio reception, with none of the usual impairments seen in analogue transmission, e.g. multi-path, white noise, etc. The main benefit of such developments is the possible addition of extra services that can be offered.

There are two major directions for the development of DTV:

(i) To deliver an increased number TV programmes of standard definition in the existing bandwidth - the approach taken by Digital Video Broadcasting (DVB) in Europe;

(ii) Broadcasting of high definition television (HDTV) programmes in the existing bandwidth - the approach taken by American Telecommunications Standard Committee (ATSC) in the United States of America and in Australia.

Both standards are based on the use of the MPEG-2 compression algorithm (the second generation source coding algorithm developed by the Moving Pictures Expert Group) with the major differences being in the types of error control coding and digital modulation schemes chosen for a particular environment. Furthermore, with the introduction of new DTV systems and services (e.g. interactive TV, video broadcasting in third generation mobile telephones, etc) these differences become more and more apparent.

The main aim of this book is to provide the reader with a comprehensive description of all error control coding and digital modulation techniques used in the DTV. The book commands a potentially wide audience including both undergraduate and postgraduate students, broadcasting and communication engineers, researchers, marketing managers, regulatory bodies, governmental organisations and standardisation institutions of the digital television industry. It also will be useful to everybody who wishes to discover the advantages of digital TV and the power of channel coding in delivering these advantages. The classroom market will be such that the book can show not only the theory of the subject at hand but connects all the theory to achieved practical aspects and provides the reader with a glimpse at what the future holds in the subject. Immediately the book will provide the reader with the knowledge that all the theory of communications can be brought together in a concise and clear manner proving it to be one of the most important fields in society today.

In the book the reader will find the relevant elements from the expansive theory of channel coding and how the transmission environment dictates the choice of error control coding and digital modulation schemes. These will be presented in such a way that both the "mathematical integrity" and "understanding for engineers" will be combined in a complete form and supported by a number of practical examples. In addition, the book will contain descriptions of the existing standards, and will provide a valuable source of corresponding references.

As DTV is a very young and dynamic field of communications engineering, the book also will include the description of the latest techniques which could find application in the future DTV systems. These include the concept of soft-in-soft-out decoding, turbo-coding and cross-correlated quadrature modulation to name but a few, which will have a prominent future in improving efficiency of the future DTV systems.

The digital broadcasting standards provide the state-of-the-art at the current time and shall remain in use for many years. The expertise gained in developing one scheme is used and further developed for new schemes and standards. Thus in the future the book will provide a basis for probably all DTV schemes in existence, thus proving its warrant for inclusion in any relevant person's library.

The organisation of the book is summarised below:

1. INTRODUCTION: In this Chapter we will present the brief history of DTV and describe the regional differences and the correspondingly the relevant standards. This will include the European digital video broadcasting (DVB) project, North American HDTV and Japanese multiple sub-Nyquist sampling encoding (MUSE) and (enhanced television) systems (EDTV).

2. BASIC PRINCIPLES OF DIGITAL TV BROADCASTING: This Chapter will explain the rationale for developing DTV, explain the role of the MPEG-2 compression algorithm and describe the necessity for using different error control coding and digital modulation formats for different environments. We also briefly discuss the role of conditional access (CA) in protecting broadcaster's

revenues. Finally we will look at the wider feature of the interactive digital TV, what is in use today and what the viewer can expect to obtain from such a facility.

3. MODULATION TECHNIQUES IN DIGITAL TV BROADCASTING: In this Chapter we will present the theoretical background of the major digital modulation formats and describe all the modulation techniques used in the DTV. These include the following:

(i) phase shift keying (PSK) modulation used in the digital satellite broadcasting (QPSK) and digital satellite news gathering (8PSK); (ii) quadrature amplitude modulation (QAM), used in the cable systems (MQAM, M=16,64,...,256), digital satellite news gathering (16QAM) and terrestrial broadcasting (hierarchical QAM); (iii) vestigial side band modulation (VSB) used in ATSC HDTV system; (iv) orthogonal frequency division multiplexing (OFDM) used in the terrestrial broadcasting systems.

In addition, the Chapter will describe the corresponding demodulation, and symbol and carrier recovery techniques.

4. ERROR CONTROL CODING IN DIGITAL TV BROADCASTING: This Chapter will include theoretical background of the error control coding, presented in a thorough but comprehensible manner. It will describe the basic principles of block and convolutional codes and thoroughly explain the two major error control codes used in DTV:

(i) Reed-Solomon block codes and (ii) convolutional codes.

Furthermore, the state-of-the art decoding algorithms for these codes will be presented, supported by a number of worked examples. Finally, the principles of trellis coded modulation (TCM) and pragmatic trellis coded modulation (PTCM) will be shown to further evolve the use of error control codes in channel coding.

5. EXISTING STANDARDS IN DIGITAL TV BROADCASTING: In this Chapter we will present the existing DTV standards on channel coding and modulation and show how the channel environment dictates the use of the particular schemes. This will include the full range of DVB standards (DVB-S, DVB-T, DVB-C, DSNG, LMDS, MMDS, RCS, RTS), the ATSC HDTV standard, the Japanese standard and the DOCSIS cable standard used in the USA.

6. FUTURE TRENDS IN DIGITAL TELEVISION In this Chapter we show that despite all the advantages of modern DTV, the efficiency of the current systems is far from the theoretical Shannon limit. We present an easy way to calculate the Shannon efficiency of any DTV system and compare the efficiencies of the different systems we have met up to this point. After identifying the gap between the theoretical limit and the efficiency of the current standards, we suggest a number of different techniques that could be used to close the gap. This includes not only a feasibility study of some particular techniques,

for example turbo-codes and the soft-output Viterbi algorithm (SOVA), but also recommendations to avoid the limitation of these techniques, for example the error floor of turbo-codes.

7. REFERENCES: In this Section we will present an up to date bibliography which will include most of the important and relevant papers in the field.

Gordon Drury
Garik Markarian
Keith Pickavance

Chapter 1

AN INTRODUCTION TO TELEVISION

1.   TELEVISION HISTORY

Television, or seeing at a distance, was visualised with great prescience in the mid-19th century by Albert Robida, a French illustrator. In 1869 the Illustrated London News journal published a drawing by Robida that showed a Gentleman at home, relaxed on his chaise longue, smoking contentedly and enjoying what was declared to be a performance of Faust. For sure this was a worthy, uplifting subject for such a viewer, but the picture shown appears more like a chorus line, dressed in the French theatre style of the time. It is clear that the illustrator had divined immediately what appeal the future technology would have to the viewer and had captured the original couch potato, viewing material more salacious than edifying. Prescient indeed!

Practical "Television" was foreseen quite soon after this, in the "wired" age of the telegraph and the telephone, by Carey (1875) in the USA and Bidwell (1881) in the UK. Lucas, working in England in 1882, realised that some form of image scanning would be required to capture a scene. The dominant technology of the day at this time was mechanical and the first serious practical work in TV derived from the experiments of Nipkow and his rotating disc scanner patented 1884/5. Revolving mirrors for mechanical scanning were proposed by Weiller in 1889. Electronic elements came in 1897 with Braun's CRT (with Fleming's magnetic deflection of 1896) and then in 1907 with Rosing and his hybrid ideas for mechanical image scanning but CRT display. Among the pioneers of purely electronic television was A A Campbell-Swinton in the UK who, around 1910, was advocating CRT technology for both image scanning and display however the technology of the time was not capable of realising his vision.

The essence of practical television, at least to engineers and experimenters, was, and remains, the reproduction of natural scenes and spectacles with sufficient realism and fidelity to the original that the viewer could become engaged with the experience of viewing. Despite the many improvements and innovations in technology during the 20th century this goal has remained a constant challenge and engineers still strive for better quality. It has taken a long time to perfect the TV systems now commonplace around the world but the essential notions of viewing, and of course hearing, at a distance has been substantially achieved. The search for higher technical quality remains today in the quest to improve the resolution of picture detail through High Definition TV but also in the quest for 3D TV where the goal is the equivalent of Surround Sound in sound broadcasting and recording. However the viewer is not necessarily interested primarily in technical picture or sound quality as such. Viewers and broadcasters see the value of TV in the "content" – the programmes – and, from a commercial viewpoint, broadcasters and the owners of content, including the film and cinema industry, see TV as a business and the viewer as a consumer.

This was clearly illustrated sixty years later than Robida's vision, in 1929, when the cinema was in full flow with the "Talkies" and television was close to becoming a practical reality. William Paley, Chief Executive of the Columbia Broadcasting System (CBS), one of the largest sound broadcasters in America, envisaged the future of television as follows:

> I visualise the world series baseball games, automobile and horse races transmitted the instant they occur on supersized natural colour, stereoscopic screens

Experience had tempered expectations by 1929 but there is some commonality between Robida's vision and that of Paley. It is interesting to observe that both visions have been realised to a large extent in terms of the content of modern television including, as it does, sport and theatre and drama of all kinds and not a little salacity. Paley's vision, another 70 years further on, is almost exact except for the "supersized" and "stereoscopic" screens. It is arguable that the former has been partly achieved if the size of the typical 1929 screen is taken into account. The screen of the first television receivers used in the 1930s was a round faced CRT with a radius of perhaps 15-20 centimetres whereas today's television set has a rectangular CRT screen typically three to four times larger in diagonal. Perhaps it only remains for engineers to perfect a 3D television system to complete the vision.

What Paley perhaps failed to state was that broadcasting, as the means of delivery, was a user of radio spectrum and would be limited in growth unless efficient exploitation of bandwidth was achieved. Despite the advances in picture quality over the years even today the analogue modulation schemes in use for delivery of TV to the vast majority of the world's viewers are little better than those available in the 1930s – mostly simple but inefficient Amplitude Modulation. As the later Chapters of this book will show, digital modulation techniques already enable an explosive increase in the number of TV channels delivered by means of radio spectrum. This will gather pace in the future and will release scarce spectrum for other uses that may include advanced forms of TV such as HDTV and 3D TV.

Much of the fundamental understanding of modern television systems was developed in the 1930s in the USA by ATT, RCA, General Electric, Westinghouse and Farnsworth [1, 2, 3, 4, 5, 6, 7, 8, 9, 10, 11] and in the UK by EMI [12, 13, 14, 15, 16] and Marconi [17]. There were also workers in Germany, Russia and Japan that contributed towards progress but these were less well and widely publicised [13].

2.   THE HISTORY OF RADIO

In the first half of the 19th century Faraday had established the links between electricity and magnetism and had made practical demonstrations of electro-magnetism. Maxwell (1864) had derived the basic mathematics that described electromagnetic phenomena and thereby predicted the existence of radio waves that could travel anywhere, apparently unaided by any physical means. After Hertz (1888) and Marconi (1901) [17] had pioneered the practicality of radio technology, it was but a small step to envisage that this new technology would be able to carry speech and music and facsimile still images, just as the telephone and telegraph wires did already. The emerging technology that would later become television would surely embrace radio as a means of distributing real-time moving images with accompanying sound.

Long before practical radio technology emerged electromagnetism had enabled the development of cable or "wired" telegraph transmissions [18, 19, 20]. The idea of communicating by "cable" entered the public consciousness and has remained there ever since. The rapid expansion of railway networks all over the world during the 19th century was greatly facilitated by the telegraph [21] and civil communication networks were also built, for example, in the USA by private companies such as Western Union. In Europe the great Victorian engineer Isambard Kingdom Brunel designed steam ships that, among other things, laid transatlantic telegraph cables [22]. Similarly, the telephone system was developed for direct speech communications also using wired means. International links were also built and led to the need for standardisation through the International Telecommunications Union (ITU) that still governs these matters to this day. The Union has always dealt with the "wired" (ITU-T) and "radio" (ITU-R) domains separately because, historically, the two areas developed in competition but also at different times and in different ways. Today, the ITU forms part of the United Nations but originally was formed from the International Consultative Committee for Telegraph and Telephone (CCITT, from its equivalent French name) and the International Consultative Committee for Radio (CCIR).

One major technical requirement for both wired and radio transmission systems was the need for a coding scheme, what today would be called "Channel

Coding", to match the properties and capabilities of the channel medium – the cable or radio channel – to the message data. Several successful telegraph codes were introduced but the most successful by far has been the eponymous scheme proposed and developed by SFB Morse between 1835 and 1844. Morse's code came to dominate diplomatic and commercial traffic during the 19th and early 20th centuries. The design of the code illustrated the solution of many of the fundamental problems facing communications engineers and has remained as an example to students of efficient communications, information theory and channel coding ever since.

The interesting aspect of the telegraph and the Morse code is that they use a discrete coding system with a finite alphabet of variable length symbols that was primarily designed to deal with text messages. This is an essentially "digital" system and incorporated statistical features that matched the Morse symbols' length to the frequency of occurrence of their equivalent letters in text. Huffman and others were to redevelop this coding feature in the mid-20th century. Indeed the development and improvement of telegraph technology in the early part of the 20th century [23] was to be important later when digital techniques returned to prominence in the 1950s and 1960s. Telephone speech, essentially an analogue signal, required a relatively continuous and unbounded channel and so led to the emergence of types of modulation systems that were optimised for conveying analogue signals that included speech, music and, eventually, TV. Broadcasters have been, from the 1920s to this day, constrained to use Amplitude or Frequency modulation with analogue baseband signal formats. There is no record of anyone in the pioneering days of TV experimenting with a discrete real-time moving image coding format – that is, TV - that could employ Morse type telegraph code. However, there is evidence [24] of still image transmission – newspaper pictures – using these techniques that anticipated Pulse Code Modulation (PCM) to some extent and succeeded as the BARTLANE facsimile system used commercially on transatlantic telegraph cables from 1922.

War, as ever, is a great spur to technological development and World War 1 saw the rapid development of radio for military communications in both speech and "wireless" telegraphy. Morse coded radio transmissions became the norm and the conflict also saw the emergence of strong encryption technologies that ensured that the insecure nature of broadcast radio was protected. The results of this were to be significant for broadcasting many decades later as will be described below. Morse coded radio transmissions were also a major feature of World War 2 during which the relevant radio and signal coding technologies, including message encryption, received another forward impulse. After World War 1 the new phenomenon of "broadcasting" emerged as a commercial prospect, powered by the technical developments of the conflict. As the 20th century progressed, broadcasting, firstly of sound alone but later of television, was to become very powerful tool that had social, political, commercial and technological consequences and, although it has been challenged, remains firmly in place to this day.

There emerged a natural and synergistic link between radio technology and sound and television broadcasting. Further, it was soon recognised that the radio spectrum was a finite and scarce natural resource and, because the early technology required expert management and control, the broadcasting business became a national asset that was best exploited for the benefit of the community as a whole. This introduced regulation by governments who, in many cases, set up specially commissioned organisations to manage broadcasting in all its aspects. The scarcity of the spectrum has been a constant factor influencing the broadcasting policy of governments. However, at the end of the 1980s it was shown that the old rules were outdated and spectrum does not have to be managed in the same way as it once was [25]. One major reason for this was the rapid emergence of highly efficient digital picture and sound coding algorithms together with very efficient digital modulation and transmission techniques that are very close to exploiting spectrum resources to the ultimate degree. New technologies and new regulatory policies that derive from them are now energising a potentially massive growth in the number of television channels and methods of their delivery to the home. Now radio technologies are not the only way that television services can be supported and so commercial operators of radio spectrum have to become very much more competitive in their approach. One key factor in the successful management of such a business is the use of efficient modulation and transmission systems and technology. The later chapters of this volume indicate the current state of the art in this area and also indicate future possibilities for sophisticated and advanced schemes to exploit the radio spectrum for television services.

3.   TELEVISION STANDARDS

3.1   INTRODUCTION

The BBC began the world's first regular high definition television service on November 2nd 1936 [13, 14, 15, 16] using an all-electronic system. This system was black and white only and used 405-line interlaced raster scanning with a 25 Hz picture rate and was transmitted in Band I spectrum at about 50 MHz. So durable was this standard that it was only discontinued in the UK at the end of 1984.

In the USA, before World War 2, experiments were carried out [2, 6, 7, 8, 11] and afterwards a system using a scan of 525 lines with a 30 Hz [1] picture rate was introduced. In the early 1950s colour was introduced (see below) and the eponymous National Television System Committee (NTSC) system is still used in the USA, Canada, Central and South America, Japan, Korea, the Philippines and other parts of the Far East.

In the late 1940's European counties began adopting systems based on a 625 line raster. During the 1950's this spread from Sweden to Gernamy, Italy and the Netherlands. In the late 1950's the CCIR collated all the variant systems then in use and published their specifications. However the CCIR was not able to choose a single world standard from these and the television business has had similar difficulties ever since. The 625 line system, together with colour (see below), added in the late 1960s using the Phase Alternate Line (PAL) scheme [26, 27], is also still in use today in Europe, including Scandinavia, China, Australasia, the Indian sub-continent and most of the Middle East. The differences in these standards caused the need for standards-conversion equipment, realised at first using analogue methods but later with digital techniques [28], to be used for programme interchange between the regions of the world, particularly when this is 'live' via satellite or submarine cable. In the UK and elsewhere scanning standards conversion was also needed to supply the transmitter networks with the two different video formats. Rather than support two networks, only the new 625-line video format was distributed to the transmitters and a local digital line rate converter was used to derive the old format for transmission [29].

3.2   COLOUR – NTSC, PAL AND SECAM

Colour television is an illusion; in practice, most of the colours found in nature can be approximated by a colour reproduction system developed and codified during the 1920s and 1930s by the Commité International d'Eclairage (CIE). The CIE basis was studied in the 1950s by the National Television Standard Committee (NTSC) during the development of colour television in the USA and NTSC was adopted by the Federal Communications Commission (FCC) in December 1953. Over a decade later, in the mid-1960s, Europe adopted the same basic scheme for colour television in which the camera resolves incoming light into its component Red, Green and Blue parts [27].

True natural colours are defined by the wavelengths of the electro-magnetic radiation in the visible spectrum between about 400 and 800 nanometres, for example Yellow is in the region of about 450 nanometres. However, the eye will accept an appropriate mix of Red and Green light as a direct substitute for Yellow; similarly a mix of Red and Blue causes the sensation equivalent to Magenta. In order to make a coded colour signal compatible with existing monochrome receivers, it was required that the colour information be added conveniently to the existing "composite" video signal. This "composite" comprised the image brightness information and the additional information needed by the receiver to display a picture synchronised with the source camera. In the case of colour, the 'brightness' signal is generated and transmitted as if it had come from a 'black and white' camera. Fortunately, another illusion can be invoked since a mix of Red ($\approx 30\%$), Green ($\approx 60\%$) and Blue ($\approx 10\%$) light appears colourless. Thus at the colour coding stage at the studio a brightness or 'luminance' signal is made together with two other colouring signals - chrominance - by taking appropriate different proportions of the Red, Green and Blue signals. These three intermediate luminance and chrominance signals are called 'components' and are stages on the way to generating the colour television signals most commonly encountered.

In all of Western Europe but France the colour coding standard is Phase Alternate Line (PAL); in France the SEquential Couleur A Memoire (SECAM) system is used. Although most of the fundamentals of the two systems are the same, one cannot be received completely on a receiver made for the other. The CCIR Report 624 describes all the essential detailed features of the standards used world-wide; the UK version of the PAL video signal is known as PAL-I [26, 27] and differs only in detail from other forms of PAL video. Differences between versions of PAL (and for other formats) emerge when the complete broadcast signal itself is examined; an example is the sound carrier frequency offset from the vision carrier.

The nomenclature should be noted here: the term 'video' is used for the baseband video but 'vision' is used for the component of the modulated RF signal which is caused by the video. The sound carrier frequency offset is fixed by the bandwidth chosen for the vision; in the UK this offset is 6 MHz whereas in continental Europe it is 5.5 MHz and in China it is 6.5 MHz. In the PAL standard the colour information is conveyed by amplitude and phase modulation of an in-band sub-carrier added to the luminance thereby producing a Frequency Division Multiplex of the separate parts. This scheme needs a phase reference that is sent during the TV line blanking periods as a burst of 10 cycles of un- modulated sub-carrier. SECAM operates in a similar way but uses frequency modulation instead of phase modulation. The single video waveform that comprises all the information needed by a receiver to recover and display a colour picture is called the "colour composite" or composite for short.

The signals described above are those that have been in wide use for many years. More recently, newer television broadcasting systems have emerged but have not yet achieved wide usage. The following provides a brief description of some of the new television formats developed over the last decade or so; its purpose is to give an awareness of these standards for completeness.

3.3 THE MAC FAMILY

The MAC system, first proposed in 1981 [30], was introduced as a hybrid analogue video/digital audio format that would be component-based and avoid some of the problems associated with the processing of PAL and SECAM signals. It was adopted in Europe as a standard for Satellite television in 1983 and found uses in other parts of the broadcasting chain [31, 32]. The MAC system was stimulated directly by the coming of Direct Broadcasting by Satellite (DBS) in the 1970s and 1980s. It was recognised very early in the evolution of the MAC concept that its new start made it an ideal vehicle to carry enhanced and higher definition television pictures. These pictures would be carried in a processed form, making full and efficient use of a satellite FM channel, and with receiver processing to restore a high quality picture for display on a suitable device. Unlike PAL, which uses a Frequency Division Multiplex format (see above), MAC takes a Time Division Multiplex form which has advantages for basic picture quality and resilience to satellite channel noise. In MAC there is no colour sub-carrier and the line and field synchronisation is done digitally.

It should be noted that despite the commercial difficulties [33] that caused the demise of the official UK satellite broadcaster, British Satellite Broadcasting (BSB), and the cessation of MAC transmissions in 1990, the MAC system is still in use and the satellites constructed for BSB have been re-deployed in Scandinavia.

3.4   W-MAC

Wide screen or W-MAC is simply a slightly modified form of MAC that deals with the wide screen transmissions using an aspect ratio of 16:9 [34].

3.5   HD-MAC

The HD-MAC system is a method of compatible HDTV delivery such that normal MAC receivers can operate satisfactorily with an HD-MAC input whilst specially equipped receivers with the decoder circuits installed and a 16:9 screen can obtain the full benefit of HDTV at 1250 lines resolution. The MAC [35, 36, 37] and HD-MAC [38] standards were developed for use in Europe and, although expressed as an analogue signal format, the majority of the signal processes were realised digitally [39]. The original plan was to make the system available by 1995 but this was curtailed due to practical and economic considerations.

3.6   MUSE

The Japanese HDTV proposal chose 1125 lines with a 60 Hz field rate. By using the same signal processing techniques as were proposed for HD-MAC, the system known as MUSE was developed [40] for the 1125 line HDTV environment and has actually been in service (a few hours a day) in Japan for a number of years.

3.7   ENHANCED PAL (E-PAL)

As a result of the emergence of opportunities for satellite broadcasting in the late 1970s and early 1980s two families of solution were proposed. One, the MAC system, has been mentioned above and the second was a form of modified PAL that was intended to solve some of the same problems that MAC was meant to remove. A scheme known as E-PAL was proposed [41] and was compared in tests with MAC during the search for a suitable format for use in satellite broadcasting systems.

3.8   PALPLUS

As a further result of the MAC proposals for higher quality resolution standards for satellite television, terrestrial broadcasters gave serious thought to the new competitive environment in their industry in the wake of re-regulation where market forces were being allowed to have their natural effects. Whilst not attempting to produce HDTV standards of quality, which some believed was neither practical nor necessary, some terrestrial operators supported the development of Enhanced PAL to place themselves better to compete with any threat from satellites or cable.

The main features of the PAL enhancement processes are:

- Wider aspect ratio, but with acceptable effects on the normal 4:3 screen,

- Reduced levels of coding artifacts such as cross-colour etc.,

- Better sound system,

- Mitigation of propagation and multipath effects eg echo-cancellation,

- Improved resolution,

- Compatibility with existing receivers.

The PALPlus system, based on the above features, has been standardised [42] and services are being transmitted by a number of European broadcasters. The marketing of the receivers has been tied with that of Wide Screen TV sets and, in some countries in Europe, this has been successful.

3.9   HIGH DEFINITION TV

The history of television is indeed the history of "High Definition" because the early pioneers set themselves the goal of television resolution that rivaled the cinema (ie 35 mm film) and stretched the technology of the day to its practical limits [43, 44, 45]. Then, as now, "HDTV" was a term which meant all things to all men and there was some debate and considerable experimentation to determine the best practical set of parameters to provide acceptable high resolution pictures both in programme making and in the home. From a technical viewpoint, HDTV is about image resolution, where the number of lines is a primary parameter from which others are derived, and resultant picture quality must be in some sense 'better' than what is already available [46]. The most important features required of HDTV are:

Better Resolution

- Spatial

- Temporal, ie no visible motion artifacts

- Removal of scan and colour coding artifacts

Wider Aspect Ratio

- 16:9 rather than 4:3

Improvements in transmission performance

- Signal-noise ratio

- Linearity etc. for analogue schemes and

- Low bit error rate and quantisation or coding defects in the case of digital systems so that the additional signal resolution is not masked by transmission defects.

Compatibility with existing context

- Ready conversion to/from existing scanning standards

- Awareness of convergence of technologies, eg the computer industry

- Realistic commercial introduction scenarios

- Spectrum efficiency

The CCIR, when it began studying HDTV in the mid-1980s, defined HDTV systems as those with more than 1000 scanning lines. At that time the world broadcasting community was seeking a single world programme production standard, a very laudable and worthy objective, but one fraught with political and practical difficulties. There were several proposals for HDTV standards and these included:

1250 lines (Europe),
1125 lines (Japan),
1050 lines (USA).

These proposals were hotly debated until the end of the 1980s when it was realised that early progress was not going to be made. A single standard for use in television production has since been agreed using 1080 active TV lines.

In the early 1990s the rapid emergence of conventional definition, multi-channel digital television systems changed the direction of standards and service planning work such that, for the time being, the majority of operationally deployed digital television systems are of standard definition. The issue of HDTV standards abated somewhat during the mid 1990s, except in the USA where there has been a prolonged process to define a terrestrial HDTV broadcasting system that was launched in late 1998. Similar interest in HDTV has been actively pursued in Korea and Taiwan, using the ATSC system, and in Australia a HDTV service is planned to start in 2001. More recently, some progress has been made with HDTV issues and, although the economics and standards issues are still hotly debated, the technologies have improved, particularly in the area of affordable HDTV display devices. Perhaps, when the conventional digital systems now being used have become well established, broadcasters will return to the issue of HDTV and reconsider the means to acquire HDTV programmes and to support their transmission.

4.   EMERGENCE OF DTV

4.1   BROADCASTING

Digital Television in the broadcasting context is not a new or even recent phenomenon [47]. In the immediate post-World War 2 period interest in applications of newly developed technologies to non-military areas led to renewed activity in television. The theoretical work of Nyquist, Shannon and others between the 1920s and the 1950s [23, 48, 49, 50] had laid foundations and the invention of Pulse Code Modulation in 1938 [51] led to theoretical and experimental work on television signal processing and on digital transmission systems [52, 53, 54], including early ideas about bandwidth compression [55, 56, 57, 58, 59]. From the early to mid – 1960's and through the 1970s, the coding of various signals, including audio and TV, had been studied with a view to their coding and transmission [60, 61, 62, 63, 64, 65, 66, 67, 68, 69, 70, 71, 72, 73, 74], particularly via digital telecommunications transmission networks for which standards were developed by the broadcasters in co-operation with the ITU [75, 76, 77, 78]. The applications that were envisaged for these standards involved the interchange of TV signals in a professional context and would not be appropriate for use in the context of delivery to the consumer.

The digital telecommunications network operators of the 1960s, 70s and 80s were deploying digital technologies for telephony speech transmission and switching by overlaying new signal formats on the existing physical media such as twisted pair and co-axial copper cables and microwave radio relay systems. The digital schemes shared these media with the existing analogue systems. The main impetus was economic, not merely technical, and it was perceived that switching cost in particular would be reduced with digital methods. Standards for digital telephony were developed from the 1960s and hierarchies of bit rates were agreed to enable wide-band high bit rate multiplexes of many telephone channels to be constructed for trunk networks. The hierarchies [77, 78] differed in the North American, European and Japanese regions for local and other reasons. The basic bit rate for a single telephone channel was chosen to be 64 Kbit/s and 2000 of these in a multiplex would therefore need a rate of 128 Mbit/s. With network management and link synchronisation data included as an overhead, the rate chosen for the hierarchy in Europe was 139.264 Mbit/s [77, 78].

High-level system synchronisation in these networks can be achieved in several ways. One way is to make the network "synchronous" where the clock is identically the same frequency everywhere. Another way is to make the network "plesiochronous" where the clocks at different network signal processing nodes are allowed to vary within defined limits and all the bit streams contain disposable time slots that are used as ballast to deal with the lack of total synchronisation between nodes. Both Synchronous Digital Hierarchies (SDH) and Plesiochronous Digital Hierarchies (PDH) are now in widespread use. Newer schemes such as Asynchronous Transfer Mode (ATM) use packet transmission techniques and are more truly asynchronous and this technique is also in widespread use. All three of these technologies have been studied by broadcasters and standards have been developed to enable TV and sound signals to be conveyed through the networks built upon them. Currently much thought is being expended on using Internet Protocol (IP) to convey broadcast television and sound in both telecommunications and Information Technology networks.

The main technical stimulus for these developments derived from the coming together of semiconductor technology and Pulse Code Modulation (PCM). In telecommunications networks these early ideas were a step on the way to the realisation of Integrated Services Digital Networks (ISDN). The ISDN concept meant, among other things, the removal of separate link and routing systems and procedures for television and telephony with an attendant reduction in cost and improvement in network flexibility for broadcasters [74]. Now, the influence of image and sound processing technologies, among others, is such that computing, telecommunications and broadcasting are converging into an integrated whole with the result that, in the future, the distinctions will disappear, not only for the image producers but also for the viewers.

It is clear, therefore, that digital television has been in an emergent state for at least the last thirty years. Because of the appearance of the right political, technical and economic conditions in the last decade, digital TV has recently become an operational fact for many commercial and public service broadcasting organisations and consumers all over the world. In addition to video and audio compression techniques there have been corresponding developments in software and in inexpensive but powerful silicon devices that make possible the complex processing necessary in the receiver. The realisation of novel and efficient modulation techniques, the main subject of this book, have also emerged to allow very efficient spectrum usage that increases the number of television channels per Hertz of bandwidth by about an order of magnitude. Typically a PAL or NTSC analogue television signal occupies 6, 7 or 8 MHz of terrestrial bandwidth, depending on the region of the world. Digital video compression techniques provide reasonable picture quality, equivalent to PAL or NTSC, at about 3 Mbit/s. Modern digital modulation systems can provide a net channel capacity in excess of 20 Mbit/s within the available channel bandwidths allocated to terrestrial broadcasting. For example, in Europe where the channel bandwidth is 8 MHz, the selected digital standard provides a net capacity up to 30 Mbit/s thus providing up to 10 times the number of television services in the same spectrum resource. For satellite systems the transponder bandwidth available, typically 36 MHz, provides a net digital capacity of about 40 Bit/s using relatively simple modulation (this enables simple receiving equipment) thus offering about 12 digital television channels in the space formerly occupied by a single analogue television signal. Such changes have radically altered the economics of television delivery to the home and have altered the long held need to ration scarce spectrum resources in the public interest.

4.2   COMPUTING

Whilst the broadcasting community was developing its approach to digital technology the computing industry was also progressing towards building competence in image processing. This was a natural development that had a wide range of applications including industrial process control, robotic vision, flight simulators, publishing and data bases. The emergence of the Personal Computer in the 1980s slowly began a process of evolution towards modern Information Technology systems that inherently include image processing. The speed and power of early machines were unable to support smoothly moving images and could not, by any criteria, be equivalent to broadcast television quality.

Time and Moore's Law ensured that the power of the PC improved very quickly and, by the early 1990s, proprietary schemes providing passable quality moving images became available. One such scheme was launched by Intel –Indeo, a cross-breed from INtel and viDEO, – that showed the potential of where the future would take personal computing. These ideas became part of the Internet and indeed made it possible and, even though it is still not able to compete in quality with broadcasting, there are clear benefits from using digital video in new ways to provide new services that will provide the impetus to improve quality. Whether the "Push" style of broadcasting, where scheduled programming is offered to the viewer who has to do little but consume it, will be augmented by the "Pull" style of the Internet, where the material to be consumed is selected consciously by the viewer, is an interesting question. In the progression of convergence among Computing, Broadcasting and Telecommunications the style of content consumption and its development is currently one of the great speculations that informs the business planning of broadcasters and internet service providers alike.

5.   THE BUSINESS OF TV
5.1   THE INFRASTRUCTURE OF BROADCASTING

The significant parts of the broadcasting chain are:

- Programmes and their production. These can be pre-prepared and purchased externally eg films, or specially commissioned and produced for television. Programmes are often produced in segments in different places over a period of time and transmitted to a studio using high quality links, or they can be 'Live'. This is the Contribution process. In North America this is called Back-Haul. This signal origination is often followed by post production where the parts of a programme are assembled and edited to the final form required by the programme makers.

- The compilation of the completed programmes into an advertised schedule and their presentation in an orderly sequence. This is often called 'Playout'. The location of the playout facilities need not coincide with the production location.

- Network transmission from the playout centre to the terrestrial transmitter, satellite up-link site or Cable Head from where the signal is transmitted. This is the first stage of the Distribution process and is known as Primary Distribution.

- The Emission or radiation of the signal from Terrestrial transmitter or Satellite. This is the second stage of distribution and known as Secondary Distribution. Technically, the satellite is only a repeater but actually does the broadcasting; transmitting to the satellite is part of the distribution process but, normally is deemed to be part of broadcasting. When satellites with on board processing capability, eg switching or multiplexing, become available then broadcasters will surely be among the early users of such techniques to enhance their services.

- An industrial infrastructure which makes available to the viewers a range of realistically priced receivers.

These stages illustrate the progress of TV signals from their origination to their viewing by the consumer. Spectrum resources are used in several of these stages, most obviously in emission, but also in contribution and primary distribution. The latter case involves professional equipment and satellite and terrestrial networks to provide a high quality service. Because the signals conveyed in contribution and distribution are likely to be processed in various ways

during production and post- production, the highest quality needs to be maintained throughout contribution and distribution transmission. This requires that
not only the picture coding schemes are of sufficient quality, but also that channel coding and modulation provide adequate protection from the vagaries of the
channel. The types of channel defects needing attention include channel symbol
errors, radio interference, jitter caused by inadequately controlled clock recovery and bit stream framing structure synchronisation. These will be recurring
issues throughout the main chapters of this book.

5.2   REGULATION

The need for regulation stems partly from the need to control access to
scarce spectrum. Regulation also arises from the need to ensure that the scarce
resource is given to people and organisations that can use it responsibly in the
public interest. On a world perspective, however, different regions have adopted
contrasting approaches; a more commercially driven regime, such as that in the
USA, or a public service regime, common, until recently, in Europe generally.
A third approach is used in some regions and countries where there is complete
state control and the broadcasting function has a more overt political purpose

Now that other methods than conventional broadcasting are available to offer
a range of services and types of programming to the public, there is not the same
need for regulation to protect and share the limited medium capacities among
the players. The following lists some of the alternative media that could be
used to provide video services:

- Terrestrial UHF/VHF

- Cable TV

- Satellite TV

- Optical Fibre and Copper Pairs used by telecommunications operators

- Traditional videotape or newer disc-based systems such as Digital Versatile
  Disc (DVD)

- Computer systems – MultiMedia with Internet links via the telephone line
  or broadband equivalents such as ISDN.

The trend towards de-regulation is gathering momentum and will encourage
the commercial exploitation of opportunities to develop these media in an environment that will be increasingly competitive. This may mean, among other
things, a need for the harmonisation of standards so that digital services can
be passed through any of these systems with minimal modification so that the
consumer's experience is enhanced and the complexities of the technology are
effectively hidden. Such changes that are implied are already happening and

the effects will be deep and significant on media companies and organisations. Whilst the processes of content origination identified above will remain fundamental, the means of their achievement will not and already there are massive changes talking place in the industry as commercial power shifts among the players.

In this change is opportunity and technology will play a significant part in enabling change especially where some of the technical issues are basic as is the case with the exploitation of radio spectrum. Efficient use of spectrum through digital modulation systems will be crucial to the ability of broadcasters to defend their business interests against the strongly emerging competition from other media especially the Internet. In the past, new broadcasting services were introduced as a result of new spectrum allocations. Even now, if a new delivery technology has been developed to an advanced stage, its use will be limited by the degree to which new spectrum is available or existing spectrum can be further exploited.

The technical aspects of spectrum allocation are of course the same everywhere but the administrative processes, criteria and controls can be and are different around the world. A complete discussion is not relevant here but it is essential to realise that regulatory matters still have a significant influence on broadcasting. Technological change can bring about regulatory change and this has been a dominant feature of recent years in Europe, largely connected with the developments in satellite delivered television. The existing regulatory regimes in Europe vary from country to country and there is an attempt by the European Commission (EC) to harmonise the rules and procedures throughout the community. This will take some time to have any effect on traditional terrestrial broadcasting but is somewhat easier to manage for satellite systems thus making it possible for these systems, for which uniform regulation is vital for obvious reasons, to develop more quickly across the community. Cable TV systems are also subject to regulation but the Internet is not, neither are tape and disc based systems.

As an example of how the regulatory process functions let us consider the current regime of regulation in UK broadcasting. It is derived from the Broadcasting Act of 1990 whose provisions are administered by the relevant departments of government. These are the Department of Trade and Industry (DTI) and the Department of Culture, Media and Sport (DCMS), formerly the Department for National Heritage (DNH) until the change of government in 1997. The DTI, through the Radio Communications Agency (RCA) is responsible for all national and international technical aspects of radio communications such as the frequency spectrum and its allocation. All internal issues relating to the administration of broadcasting such as which organisations are permitted to broadcast, their financial structures and ownership and the control of their activities is the responsibility of the DCMS. Thus it is the DCMS which sets up, usually by means of Acts of Parliament, UK broadcasting regulators such as the Independent Television Commission (ITC) and the Radio Authority (RA). The governors of the BBC are also responsible to the DCMS. The administration of Pay-TV and Conditional Access systems for digital TV services is performed by the Office of Telecommunications (OFTEL). The effect of digital technology driving convergence among the broadcasting, telecommunications and computing industries is challenging the current organisation of regulatory bodies, causing overlaps of jurisdiction to appear and leading to changes in these bodies and their inter-relationships. It is conceivable that there will be a convergence and combination of regulatory activities as time progresses and the applications and uses of digital television technology unfold; for example, one area already challenging regulatory regimes in this way is the Internet. In the wider context of the EC, the UK regulators also have some responsibilities to co-ordinate with colleagues in the Commission so that European regional regulation converges. This is likely to take some considerable time since the states of Europe are not all at the same stage of development and readiness to exploit new technologies.

5.3   COMMERCIAL MATTERS

Broadcasting is a business. It has costs and revenues and these can be managed in different ways. As has been described above, most of the world's broadcasting infrastructures are regulated and this reflects both the style and intent of government as well as the cultural environment. In the case of commercial broadcasters the business is more conventional but the public service broadcasters have to behave in specific ways and have to balance a range of issues that commercial players do not have. A mixed economy of public and commercial entities will bring some degree of conflict especially where access to key content and public events is concerned. It is clear for example that the public service remit is broad and challenging whereas the commercial operators can choose to serve only those sectors of the viewing public that will pay for their output unless the regulator constrains them to provide a broader range of service. The socio-economic grouping of viewers targeted by the advertisers that support some commercial broadcasters is a very significant element in the type and range of programmes. It has been clear from experience in the USA, and more recently in Europe, that sport and movies are key strands of reliable revenue earning potential. Whilst businesses can be focussed on, say, sport they will develop by building relationships within the sports institutions through the injection of sponsorship funds and even equity ownership in some of major players, this can be damaging both to the sport itself, by distorting its independence, and to those elements of the viewing public that cannot partake in viewing sport for economic reasons. The examples of, say, the UK football Cup Final or the Grand National horse race, possibly national assets that should be available to all without the payment of additional fees, serve to show how the public interest can be undermined by excessive commercial control. The "cherry-picking" of markets can weaken the role of the public service broadcaster who is left with only those viewers that cannot otherwise be served. The funding of the public service broadcasters must be adequate for it to compete but cannot be so inflated that it becomes detached from its original purpose. Regulation of the commercial players is therefore considered necessary in the public interest.

It is in the newer areas of Internet broadcasting that the future role of broadcasters is more problematic because they never had any overt mandate to provide services in this way. It may be argued that public service broadcasters have an implied remit to use all available media in the quest to serve the public. One benefit of such an interpretation is that the public service body provides the benchmarks of content and technical quality that encourages commercial players to respond. This works only if the public service body and the regulator are each strong enough. Many broadcasters are very active in the Internet but are criticised for spending public money on doing so. It is a dilemma that continues to exercise the minds of governments, regulators and industry players.

The delivery technologies are enablers for all broadcasters and, since broadcasting is a major spectrum user, it is essential for the success of broadcasting in its competition with the Internet that new modulation technologies are explored and deployed in a timely way. There has been no more challenging commercial era in broadcasting than that which now faces its current players and so the resources and energies of the industry should be focussed on providing opportunities that will protect the business well into the new century.

6. DIGITAL BROADCASTING STANDARDS

6.1 ADMINISTRATION OF TECHNICAL STANDARDS

When the United Nations was established the International Telecommunications Union (ITU) was absorbed into it along with its components, the existing International Consultative Committee for Radio (CCIR) and International Consultative Committee for Telephone and Telegraph (CCITT). Signatories to the UN and, hence, the ITU are bound as in treaty to the terms of reference and are obliged to respect the decisions and procedures of its committees. In a recent re-organisation the CCIR is now known officially as the ITU Radiocommunications Bureau - ITU-R - and the CCITT as the ITU Telecommunications Bureau - ITU-T.

Broadcasting, as a radio spectrum user, is administered under the CCIR/ITU-R via its Study Group 6. A further group of relevance - formerly the Mixed Committee for Television and Telephony (CMTT) - deals with the interaction of telecommunications standards with those for television as they come together when television signals are transmitted via telecommunications systems designed primarily for telephony. This group is now administered via the CCITT/ITU-T as its Study Group 9. Within the ITU there are other groups of relevance to broadcasters using satellites and these include the International Frequency Registration Board (IFRB) and the World Administrative Radio Conference (WARC). The results of these conferences find their way into the Radio Regulations which provide separately for three regions of the world: Region 1 - Europe, Africa and the Middle East, Region 2 - The Americas and Region 3 - The Far East, China and Australasia.

In Europe, the European Telecommunications Standards Institute (ETSI) has the task of producing European standards for telecommunications systems and this includes broadcasting by means of a formal liaison with the European Broadcasting Union (EBU). There has, for the last several years, been considerable activity focussed on progressing new digital broadcasting standards in ETSI. A pan-industry group called the Digital Video Broadcasting (DVB) project was formed in Europe with strong representation from both technical and commercial interests and expert groups have discussed the several aspects of a set of new standards for digital broadcasting.

In the USA, some of the official standards groups contributing to broadcasting activities are the American National Standards Institute (ANSI), the Institution of Electrical and Electronics Engineers (IEEE) and the Society of Motion Picture and Television Engineers (SMPTE) and these have been in existence for many years. A very successful commercial and proprietary digital technology was deployed in the early 1990s by DirecTv [79], a satellite broadcasting system, but, more recently, the use of DVB standards [80] by newer satellite broadcasters has also occurred. Terrestrial broadcasters in the USA have been working towards a digital system to carry High Definition TV since the late 1980s and such a system – the Advanced Television System Committee (ATSC) [81] - was launched in late 1998.

Organisations like DVB, ATSC and others have come into being as a result of the lack of response of traditional standards groups to the rapid development of new digital coding and transmission technologies. Indeed, some existing standards groups have proved more capable of responding to this challenge than others - for example, the ISO/IEC joint work in the Moving Picture Experts Group (MPEG - see Chapter 2). The ATM Forum (ATMF) has appeared to develop and define Asynchronous Transfer Mode (ATM), a fast packet switching and transmission technology. The Digital Audio- VIsual Council (DAVIC) was convened to exploit all these new technologies and define operational protocols and procedures to permit interactive services such as Video on Demand to be supplied in a unified way over a number of media. Similarly the Internet has caused the convening of the Internet Engineering Task Force (IETF). Its objectives include the harmonisation and further development of the many aspects of the network, such as interoperability, in order to make it easier to use and to address the demands made upon it by a wide range of services. DAVIC and IETF are not standards bodies, neither is the ATMF, and these informal organisations derive their authority from their support by all the important elements of the communication industry. Their objective is to prepare specifications that then become accepted practice and are then adopted and administered by legitimate standards bodies, perhaps on a world or, at least, a regional basis. This, despite the speed with which these groups work, seems to be succeeding and could be a future pattern where standards groups are formed as required when a particular technology reaches sufficient maturity that it needs development.

6.2   EUROPEAN APPROACH – DVB

During the 1980s European developments for new television systems were concentrated on attempts to define standards for HDTV (see above). Opportunities to use satellites as the means to transmit new broadcast services were stimulated by the World Administrative Radio Conference (WARC) 1977 plan for Europe[2], giving a very rare opportunity to use new technical standards in the newly created medium. The MAC/HD-MAC system described above was designed specifically for this application and used a simple form of Phase Shift Keying (PSK) as its modulation scheme.

However, the lack of commercial success caused a hiatus that was eventually filled by proposals for the use of purely digital techniques. The experience of the late 1980s had taught the European broadcasters and industrial companies that perhaps a new approach was needed to prepare and define standards for a new era. In the USA (see below) the Advanced Television Standards Committee (ATSC) had been established to develop a new terrestrial broadcasting standard and was inviting proposals from a number of consortia. These were primarily analogue hybrids taking their lead from a desire to find a backward-compatible scheme that would not disrupt existing analogue NTSC services. These same methods were also studied in Europe at this time.

During the late 1980s there was also development taking place that established the Moving Picture Expert Group (MPEG) [82] and its sister group the Joint Photographic Expert Group (JPEG) [83]. These were interested in digital picture coding schemes that would reduce the bit rate required to transmit pictures within Information Technology systems and did not at this time consider broadcast applications. European researchers grasped MPEG with enthusiasm and were soon extending its initial objectives towards uses in mainstream broadcasting. By 1990 it was clear that this work could produce useful results. This prospect, together with the failure of the MAC scheme, caused European agencies and companies begin considering means by which this new technology could be developed successfully. It was recognised that a new approach was required mainly because it was expected that digital technology would offer radically new opportunities and synergies with computing and telecommunications systems.

The European Commission (EC) shared this thinking and was supportive of a new approach. The result was consultations between the EC, the broadcasters through the European Broadcasting Union (EBU), the consumer electronics industry through the European Association of Consumer Electronics Manufacturers (EACEM), and the standards forum CENELEC and the European Telecommunications Standards Institute (ETSI). The European Launch Group (ELG) was set up in 1991 to define the means whereby these bodies would work together to establish a new regime of television standards for Europe. By 1993 the Digital Video Broadcasting (DVB) project had been established and provided with a Memorandum of Understanding, a management structure and channels of communication with EBU and CENELEC and ETSI to publish standards [84, 85]. It was also established that no technical standards would be produced unless there was a sound commercial background to require such actions.

It was agreed within the relevant DVB Commercial Module that the first application of the new digital technology would be in satellite broadcasting [86] and so a standard was commissioned from the Technical Module. This was produced quickly and published by ETSI [87] as the first of many DVB standards [80, 88, 89, 90, 91, 92, 93, 94, 95]. The philosophy of DVB was that MPEG would be adopted entirely and that DVB would not add anything to the source coding work of MPEG unless it could add value to its implementation [96]. The main objective was that DVB should continue from where MPEG had stopped and this meant defining standards for channel coding that MPEG had specifically decided not to address. In addition, DVB addressed all those aspects of the MPEG Systems layer, Part 1 of the Specification [97], that needed additional specification to allow broadcasters to use this layer in practical circumstances, for example, Service Information [94, 95] and Conditional Access (CA) [98, 99, 100] and means of enabling easy access to the receiver – the Common Interface [101] – for different CA systems and other external applications [102].

As a result of its commercial focus, DVB at first studied only those media that were operated by broadcasters before extending its work to other related media and issues. Satellite, Cable and Terrestrial broadcasting were clear candidates. Emerging and related opportunities existed in Satellite Master Antenna TV (SMATV) and Microwave Multipoint Distribution Systems (MMDS) applications. Latterly work has been concentrated on Return Channel Systems (RCS) for all these media so that Interactive broadcasting can be addressed as described in Chapter 2.

In addition a generic receiver specification, the Multimedia Home Platform (MHP) [103, 104, 80], has been developed that will ensure that in Europe at least future receiving equipment will be able to support a wide range of services presented in many ways. The inclusion in this specification of a defined Application Programme Interface (API) that can manage and control viewer access to and navigation through the services is indicative of the new era that digital broadcasting brings. The MHP is a specific response to other bodies such as telecommunications organisations, consumer electronics and cable television companies that are developing their views about the future digital home and its facilities for distributing services among members of the household. Standardisation seems crucial to enabling this vision of the future for otherwise chaos seems the only alternative. Interoperability across global networks requires an unprecedented co-operative spirit among broadcasters and others and the intense competition will ensure that this will not be easy or quick to achieve. There will be a strong economic pressure on designers of practical MHP solutions to be competitive with more proprietary schemes that avoid the costs of broad interoperability by using de facto standards, perhaps sponsored by commercial broadcasters whose business plans are specific to their needs.

The nature of the different media that could be envisaged for carrying broadcast-like signals, coded using MPEG, varies widely and so different coding is required to allow each medium to be optimised to carry digital transmissions. Because compressed video and audio are more susceptible to channel defects, selecting the appropriate channel coding is crucial for success in each case. Channel coding comprises both Modulation and Forward Error Control (FEC) and the following Chapters of this book will describe and develop these aspects further.

6.3   NORTH AMERICA

MPEG had made good progress by 1990 but had not yet made a standard. In that year the same ideas were sufficiently well developed by the General Instrument company that they made a very late submission to the ATSC [81] proposing purely digital picture and sound coding and transmission systems for a new HDTV service. The other consortia were offering complex and unwieldy analogue schemes and were caught unawares by the GI proposal. However, soon, other digital candidates were offered and these led to the development of a standard based on MPEG and the testing of its practical realisation prior to its adoption by the FCC as the current ATSC system. A number of terrestrial television stations in the USA are now transmitting regular HDTV services using the ATSC standards [81, 105, 106, 107, 108, 109, 110].

The ATSC standard was destined for use only in terrestrial broadcasting. It was begun in the late 1980s and by 1993 was well established with a draft standard that was being tested at the Advanced Television Test Centre (ATTC) a facility set up by the US industry to resolve the various elements of the competing systems. The terrestrial broadcasting frequency plan in the USA is based on a 6 MHz channel separation in both VHF and UHF bands that permits a proportionally lower bit rate – other things remaining equal – than the European 8 MHz plan. The main degradations in a digital terrestrial channel are not caused by noise but by multi-path, especially if the viewer has not invested in a good antenna system. Coverage must be close to that of the existing analogue service. There is a compromise to be struck between performance at the edge of the area of and performance well within the coverage area but impaired by the effects of urban features such as tall buildings. In the absence of multi-path conventional modulation systems work well in noise limited cases. The Nyquist rule for 6 MHz bandwidth and the Vestigial Side Band (VSB) method allows a symbol period of 93 nano-seconds equivalent to a symbol rate of about 10.7 MBaud. There is a variant that is meant for use in cable systems and it supports a doubled bit rate.

The specific modulation scheme chosen is called 8-VSB [106] and it uses a hybrid approach where traditional signal structures based on analogue television frames are grafted with new ideas. 8-VSB is a conventional multilevel single carrier system with 8 amplitude levels allowing 3 bits per symbol efficiency that gives about 32 Mbit/s gross bit rate. Using a forward error correction rate of about 2/3 (see Chapters 4 and 5) and allowing for some time for synchronisation elements in the bit stream, the net usable bit rate is about 19.3 Mbit/s. In an 8 MHz channel this should be proportionally greater.

Unlike Europe, where a co-ordinated approach to standardisation across the media had been taken, the US did not seek such harmonisation at that time. The cable industry was strong and had a significant share of the analogue market. The threat of digital competition caused a commercial reaction that did not include the thought of harmonisation. The result has been that each medium has developed its own approach to digital standards with the only common factor being the choice of MPEG for compression.

Regular digital satellite broadcasting services began over 7 years ago when DirecTv was launched [79]. In the mid-1990s there was competition from commercial operators such as Echostar, Primestar and Alphastar some, like Echostar, using the DVB standard. Now these systems have been rationalised and only Echostar and DirecTv remain. The modulation scheme is basically the same in all satellite broadcasting systems and uses Quadrature Phase Shift Keying (QPSK) that allows a good compromise between the practicality of low receiver dish size and sufficient net bit rate. The DVB system allows a 53 cm dish to deliver about 39 Mbit/s net capacity in a 36 MHz transponder.

Cable channels are not beset by the same channel defects as terrestrial systems and can be engineered to avoid many of the vagaries arising there. A modulation scheme that has noise limited performance is appropriate in cable systems and, considering that most existing cable networks were designed for analogue television signals and therefore have good signal-to-noise ratios (S/N), a single carrier scheme is acceptable. It is not necessary for digital systems to have as good a S/N as analogue systems and so greater bit rate can be supported by trading some S/N. This means that high order Quadrature Amplitude Modulation (QAM) schemes can be chosen with confidence and typically 16QAM to 256QAM is used. In a channel plan based on 6 MHz slots the Nyquist rules suggest that a symbol rate of about 11 MBaud can be supported. This, combined with a coding efficiency of 4 to 8 bits per symbol, channel quality permitting, enables a gross bit rate of between 44 and 88 Mbit/s. The use of a forward error correcting code with rate ¾ would reduce the net bit rate to between 33 and 66 Mbit/s. The US standard for cable has been developed by Cable Labs [111] and is known as Digital Over Cable Signal Interface Specification (DOCSIS) [111]. It is in service in the USA and may even find use in Europe because there are as yet few European digital cable systems in operation.

6.4   JAPAN

Japan was late in embracing digital broadcasting. Whereas its industrial companies had been active and supportive of digital standards for studio production through the ITU-R they were not so prominent in promoting similar digital standards for direct to home broadcasting. The consumer electronics industry of Japan is an immensely powerful force in commercial terms but, during the late 1980s and early 1990s, it was not a leading force in the development of digital technologies. Nevertheless, once prompted by developments in Europe and North America the Japanese consensus system engaged the issues and began to develop a Japanese digital philosophy. This led to the adoption of some DVB standards that allowed digital satellite services to be started.

The approach to the terrestrial broadcasting case has been unique to Japan. The same modulation technology as used in Europe – OFDM – has been chosen but with an emphasis on flexibility with the intent of offering both fixed and mobile services. A much more complex form of OFDM called Band Segmented OFDM – BS-OFDM – has been selected for the Japanese domestic terrestrial standard [112] and is described in more detail in Chapter 5. However, it seems unlikely that a service will be available for a number of years.

Notes

1   The picture rate was chosen initially to be exactly 30 Hz but was later changed to 30(1001/1000)=29.97003 Hz to prevent visible beating between the colour and sound carriers. The latter is set at 4.5 MHz and so the line frequency, to which the colour carrier is locked, was changed from 15.750 kHz to 4.5/286 MHz=15.734266 kHz. 4.5 MHz is a key frequency in digital television coding standards (see Chapter 2) because the European line frequency is 4.5 MHz/288.

2   The introduction of DBS in Europe had been planned on a technical level for several years during the 1970s. The CCIR had, in 1977, agreed a plan that allowed each European country to have spectral and orbit allocations. It was not until 1987 that the 5 UK channels were considered to be commercially viable and, hence, given regulatory approval. As a result of the commercial failure of British Satellite Broadcasting (BSB) in the UK [33], the only truly commercial satellite broadcaster in Europe at the time using DBS frequencies, there is now no significant activity based on the plan. The spectrum portion, 11.7 - 12.5 GHz, designated for the use of MAC, has been re-planned for digital applications and has thereby increased the number of RF channels available.

Chapter 2

BASIC PRINCIPLES OF DIGITAL TV BROADCASTING

1.   WHY DIGITAL

The decades after World War 2 saw a rapid and widespread development of computing technology and Pulse Code Modulation (PCM). Mostly for military purposes, digital computer systems using binary code soon became dominant and, following the introduction of integrated semiconductor logic circuits in the late 1950s, the cost reduced sufficiently to attract manufacturers towards commercial applications. In the 1960s telecommunication system operators were active in applying digital techniques to telephony [113]. In the mid-1960s broadcasters also began to experiment with digital coding of sound and then, later, television signals with a view to exploiting the advantages of the new format [51, 52, 53, 54, 63, 64, 65, 68, 69, 70, 72, 73].

Analogue systems of signal coding and transmission are prone to a number of defects. One of the most significant is the inconvenience of storing information. By embodying the information to be transmitted or stored in waveforms any distortion of the waveform in transit is a distortion of the information. This requires that analogue channels are extremely linear. Digital representation does not suffer these defects and information transmission and storage becomes more ideal using binary arithmetic computing techniques where each processing element is a Binary Digit or "Bit". The one disadvantageous consequence is that the rate at which the information "bits" have to flow to represent a moving television image is quite high, typically hundreds of Mega-Bits per second, and thus needs a high bandwidth to convey it. The ruggedness of the digital format is its main advantage. Any damage to the bit stream during transmission, for example bit errors, can be managed by means of error control codes. The discussion and description of such codes are major components of this book.

For broadcasters the ability to avoid analogue defects was attractive in reducing costs and improving the quality of the pictures reaching the viewer. The tendency of analogue system parameters to drift from their settings over time, requiring frequent maintenance, and the investing of the image quality in the relatively indestructible numbers of the digital format are the main reasons for the move.

Once in the digital format, there is no significant difference between broadcasters' signals and, say, multiplexed digital telephony or computer data signals. This allows the use of technologies developed in other areas such as digital computing to be adopted in broadcasting with the attendant convenience and economy that such a step provides. Digital image processing and transmission systems for broadcasting draw heavily on the experience gained in other applications and use the same source and transmission coding systems and processing techniques. The special needs of broadcasting can be accommodated readily because of the flexibility of the digital format.

Analogue broadcasting systems are simple in design and do not lend themselves readily to flexible re- configuration according to specific broadcasters' needs. The simplicity is not so much an issue at the studio or during transmission where professional equipment can be more readily afforded. It is in the viewer's receiver that the main obstacles lie in expanding and diversifying television services. By enabling digital transmissions to a digital receiver system equipped with storage and a powerful processor, vastly more flexibility can be provided to broadcasters in providing viewers with a much larger range of programmes and services.

In addition to the transmission system design and its ability to deliver digitally coded programme data, there is a need to make the much more complex digital system user-friendly and flexible in its use by both the consumer and the broadcaster. A complete system, representative of that considered for digital satellite broadcasting, is illustrated by Figure 2.1 from which it can be seen that multiplexing arrangements are also required to combine the component parts of the transmission. Each bit stream formed in this way will involve many more than one television programme per transponder and will enable the flexible re-configuring of the multiplex structure to accommodate different modes of programme presentation. For professional applications, the same degree of user control is not needed and the multiplexing need not be so flexible or provide as many channels. Such schemes have already been implemented in systems that are commercially operational all over the world.

The ability to re-configure dynamically the programme multiplex from time to time and to inform the viewer of the contents of the multiplex is achieved by adding a specific data stream to the multiplex that contains the appropriate information. This Service Information (SI) is an essential part of a practical system for broadcasters where the public is faced with complex technology.

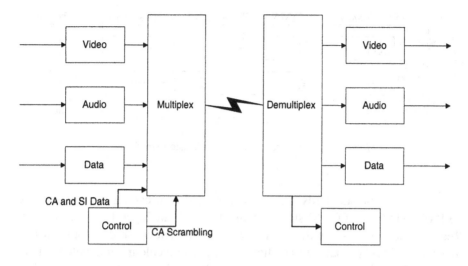

Figure 2.1.   High Level Block Diagram of Digital Broadcast System

Conditional Access (CA) data (see below) will also have to be transmitted in the multiplex.

2.   DIGITAL CODING OF EXISTING TV SIGNALS

There is an agreed international standard for digital coding of the television component signals and this can be found in ITU-R Recommendation BT-601-4. This specification defines the coding of 'Component' television signals which are the Brightness, or 'Luminance', and two Colouring, or 'Chrominance', derivatives of the original independent Red, Green and Blue signals resolved in the camera. There is an additional specification – Recommendation BT-656 - that provides for digital interfaces between equipment supporting Recommendation 601 coding parameters. These standards generate high bit rates as illustrated by Table 2.1. There are already systems using the provisions of Recommendations 601 and 656 in widespread operation on a professional basis in studios and, for contribution and distribution applications, associated video compression technology has been developed. The SMPTE has also published specifications for television signal coding and interfacing [114, 115, 116].

It should be noted that the sample frequencies are all locked to the television line scan frequency such that 13.5 MHz is 864 x PAL line rate (15.625 KHz) and 858 x NTSC line rate (15.734266KHz). This is in order to ensure that all samples of the video form a rectilinear sample structure in each complete frame and repeats in each frame. This enables ease of picture manipulation. The frequency 13.5 MHz is also a simple integer multiple of 4.5 MHz which is an important frequency in all television systems (see Chapter 1, Section 3.1).

COMPONENT	SAMPLE FREQUENCY, MHz	NUMBER OF BITS PER SAMPLE	BIT RATE Mbit/s
Luminance	13.5	8	108
Chrominance 1: [B-Y]	6.75	8	54
Chrominance 2: [R-Y]	6.75	8	54
TOTAL:			216
Luminance	13.5	10	135
Chrominance 1: [B-Y]	6.75	10	67.5
Chrominance 2: [R-Y]	6.75	10	67.5
TOTAL:			270

Table 2.1.   Bit Rates for Digital Television Standards

There are some standards [115, 116] for the digital coding of the composite PAL, NTSC or SECAM signals defined in their analogue forms by ITU-R Recommendation 624. In the early days of digital television development the first and obvious approach was to digitise the existing colour composite format. In the early period this was eminently possible and satisfactory because this format was used throughout the television broadcasting chain. Thus, in the late 1960s and early 1970s a digital video signal was envisaged as a "digital composite", constructed as a Pulse Code Modulated (PCM) version of the analogue [63, 64, 65, 68, 69, 70, 72, 73]. This digitisation process uses Nyquist's criterion to set a sampling frequency for the conversion that is more than twice the highest base-band frequency in the video that is typically 5-6 MHz for PAL and SECAM but for NTSC it reduces to 4.2 MHz. The sampling frequencies then need to be more than 8.4 MHz for NTSC and 12 MHz for PAL and SECAM.

The number of bits per sample of this video was chosen by experiment to be 8 at first. Early work [65] showed that contouring effects would be invisible at this level and quantisation noise, also the result of a finite number of bits per sample, would be comparable to the analogue noise levels normally accepted for analogue video quality criteria [117] [1]. However when it was much later realised that all-digital production facilities would need to manipulate the video, processing of this colour digital composite had to be considered with the result that 10 bits per sample was eventually chosen to allow headroom for the accumulation of digital arithmetic round-off errors.

Also studied [64] was the precise value of sampling frequency since the quantisation noise, being directly related to the video signal itself, is not always a linear additive impairment like noise but in fact is best considered as a non-linear distortion of the original video. Quantisation Distortion can be highly rich harmonically and the power distribution of its amplitude (closely related to its spectrum) can be far from Gaussian and well behaved, as normal analogue thermal noise is, and it appears as severe and highly visible patterned defects in the reconstructed images. The solution to this undesirable result of video quantisation was not to increase the number of bits per sample, although this certainly helps, but to lock the sampling frequency to a harmonic of the colour sub-carrier [64]. The effect is not to remove the quantisation noise, or even diminish it, but to mask it by causing the harmonic components of the distortion spectrum to be coincident in frequency and phase with the harmonics of the video itself. Thus PAL and NTSC composite video is processed using sampling frequencies having various harmonic relations with the colour sub- carrier, typically 2, 3 and 4 times. The SECAM system does not modulate its two colour sub- carriers in the same way (it uses FM not PM/AM) and so does not need the same degree of precision in sampling frequency value but does require it to be locked to the line frequency to allow signal processing. The 4 times sub-carrier sampling frequency gives a sample structure that is very close to a rectilinear one frame to frame that assists greatly in picture manipulation and so this, despite its greater bit rate, is preferred over the 3 times frequency. The 2 times rate has for PAL and NTSC the peculiarity that it is sub-Nyquist [69, 70] but the properties of the television signal give it special features that render it acceptable in this form. The arrangement also minimises bit rate at virtually no cost and, because it has a frame coherent sample structure, enables ease of picture manipulation especially the standards conversion process between the 625 and 525 line based video formats.

The transmission capacity needed by these schemes, shown below as Table 2.2, for PAL and Table 2.3 for NTSC is the direct product of the sampling frequency and the number of bits per sample.

SAMPLING FREQUENCY, MHz	NUMBER OF BITS PER SAMPLE	BIT RATE, Mbit/s
2 × 4.43361875	8	70.937899
3 × 4.43361875	8	106.40685
4 × 4.43361875	8	141.8758

Table 2.2.   Bit Rates for PAL Sampled at multiples of the Colour Sub-Carrier Frequency

SAMPLING FREQUENCY, MHz	NUMBER OF BITS PER SAMPLE	BIT RATE, Mbit/s
2 × 3.579545	8	57.272720
3 × 3.579545	8	85.909080
4 × 3.579545	8	114.545440

Table 2.3.   Bit Rates for NTSC Sampled at multiples of the Colour Sub-Carrier Frequency

The bit rates are high and thus demand high bandwidth of the transmission medium. There are means to reduce such bandwidths by reducing the bit rate and this can be done, either by reducing the sampling rates, or the number of bits per sample, or both. Early workers in this field called this process "Bit Rate Reduction" but the term "Compression" became more the norm after about 1990. Clearly the original high bit rate came from having taken the trouble to meet Nyquist's Criterion and reduced Quantisation Distortion to acceptable levels. If the bit rate has to be reduced, how can the quality of compressed video be maintained if sample rates and quantisation precision are reduced in the process? The answer lies in the statistical properties of the video itself that allow considerable advantage to be taken of the fact that, instantaneously, the video does not always need full Nyquist sampling and full quantisation precision. The compression process is described further below. Further detailed description of digital television techniques than is relevant here can be found in the literature [47].

3.   THE COMPRESSION PROCESS

It has been shown above that one result of using digital representation for television signals is a significant increase in demand for bandwidth. Spectrum loading with such rates is significant, even though the Carrier-to-Noise ratios (C/N) needed for digital schemes are significantly reduced compared to analogue ones. To reduce this high demand for a scarce resource it has been necessary to consider techniques to reduce the bit rates without of course affecting the quality of transmission.

Workers [55, 56, 57, 58, 59] as long ago as the 1950s recognised this fact and, even before advanced forms of compression were developed in practice, had developed simple methods such as Differential Pulse Code Modulation (DPCM) [55] where the differences between samples of a signal were coded for transmission rather than the samples themselves. Since the differences were likely to be smaller in amplitude than the full PCM sample values there is a saving in the number of bits needed to represent the signal faithfully. This differential coding arrangement is an inherent part of the MPEG algorithm (see below in Section 4). Coding like DPCM that involves calculating differences between signal sample values is also known as Predictive Coding because each sample value is effectively predicted to be the same as an appropriate preceding one in a sequence. The difference between the prediction and the actual value, presumed to be consistently small, is coded and transmitted.

There is however one clear disadvantage of DPCM and all differential schemes because any errors in transmission may corrupt the data and may completely disorient receiver synchronisation and so it will fail seriously to reconstruct the signal faithfully. One remedy is periodically to transmit full sample values so that the receiver can be reset should it be overwhelmed by errors. This feature of DPCM serves to illustrate a trait of all compression schemes and that is that they exhibit a significant susceptibility to transmission errors and so require strong systematic synchronisation methods. It seems, at first sight, counter-productive to reduce the bit rate only to have to increase it again by adding synchronisation bits and possibly error correcting codes. Fortunately the net gain in most compression systems is still advantageous. The synergistic combination of video compression with highly efficient error coding and modulation schemes is at the very centre of the success of digital broadcasting.

Compression can only function well if the picture material is substantially predictable and well behaved. When the picture material contains rapid motion and frequent cuts between different sources then compression is less effective. Also, if the statistical properties of the video are from time to time noise-like then any algorithm will face difficulties. Unfortunately normal video material is made of moving images, some motion being rapid, and of different sections edited together where there can be frequent changes of scene. If a camera is operating in low light, perhaps for artistic reasons, or a movie is being coded or the material is old and taken from the archive then an additive random noise-like disturbance will be presented to a compression algorithm. Such an algorithm will attempt to code the unpredictable noise but will consume valuable bits in doing so.

Film is inherently easier to compress than natural interlaced video, other things being equal, but film grain has the properties of noise and so raises difficulties. This is because each film frame is a progressive video frame, without interlace effects, thus improving pixel correlations that assist good compression. However, subtle complications are caused by attempts to code video that is taken from film scanning mechanisms. For example, because the film tends to weave as it passes through the telecine machine the output video image samples lose a degree of spatial coherence from frame to frame. Furthermore these effects accumulate in the film processing sequence from camera to scanned print so that there is opportunity for image skew to build up. Predictive coding that uses one frame of video samples as a prediction for the next will be presented with sample positioning errors that will appear as de-correlations and so reduce the efficiency of compression. A simple calculation shows that a total image weave of only 0.2 degrees in the movement of a 35 mm film will cause a sample displacement of one horizontal pixel. Instabilities in the film speed through the telecine, or in the pull down mechanism or flying spot in older machines, and the accumulation of such errors in the camera and during film printing can cause similar vertical sample location errors. Another simple calculation shows that accumulated speed instabilities (or image placement error in the direction of film travel) of only approximately 0.2% can result in a displacement of one vertical pixel, equivalent to a reversal of interlace. These errors are mitigated to some extent by spot size and integration effects in some telecine machines (flying spot devices whose spot is larger than an equivalent video pixel) but nevertheless these effects lose resolution. Modern machines using line array CCDs and continuous motion scanning for example, whilst providing more spatial and temporal resolution, are still somewhat prone to the effects of weave and speed variations. It is possible to correct some of these defects by detecting them and then stabilising the image in a frame store prior to compression. Some telecines can do this themselves using Motion Estimation and Compensation techniques, not only for their own failings but also some of those accumulated during film production. Electronic film production removes many of the causes of accumulated error.

Another cause of poor coding efficiency derives from using composite video input signals with an algorithm designed for component inputs. The complex combination of the colour information with the luminance is not designed for frequent decoding and re-coding. PAL, SECAM and NTSC were designed for encoding only once at the studio and decoding once at the domestic receiver. One result of using digital processing in the whole broadcast chain is to code and re-code more often. Each time this is done the overall signal quality is eroded and artifacts of an imperfect decoding operation can prevent a subsequent compressor from achieving its full potential. The highest quality composite decoding depends on digital processing with access to several fields of the image and this can remove the majority of the visible defects. The decoding process is one of filtering [118] that is close to perfection for still images but loses its effect when there is movement. The filters have to be designed as a balanced compromise between the need allow true movement to remain fluid and colour separation to be as complete as possible. In a practical decoder another defect is the leakage of a small amount of the sub-carrier to the output and this again causes the compressor to waste bits. It is necessary to specify the amount of leakage permissible to be less than -45dB with respect to the video signal for this not to be a major issue.

To mitigate all these effects it is advantageous to precede the main algorithm in a compressor with a pre-processing stage that removes random noise and attempts to reduce the effects of film weave and, if necessary, the effects of composite signal decoding. The result of such a pre-processor can be that it adds delay to the whole compression chain but, by integrating elements of the pre-processor within the compressor, any delay can be minimised by sharing frame stores. Pre-processing also adds cost to the encoding product and so vendors have taken steps to include some elements to maintain competitive performance but leave others to the operator's choice where specific conditions can determine whether external pre-processing is more relevant.

4.   ISO/IEC MPEG-1 AND -2

Compression technology has been in use for professional applications in broadcasting for some time. ITU standards for compression [74, 75, 76] exist for transmission networks. However, within the studio proprietary compression systems have emerged for applications in video production and storage. One such is the Digital Video (DV) system promoted by the Matsushita Corporation of Japan trading as Panasonic. This scheme retains quality by keeping bit rates reasonably high and by not exploiting temporal redundancy. In a production environment editing at frame boundaries is important and some compression systems compromise this possibility. The Sony Corporation of Japan has promoted a system known as SX that also maintains reasonably high bit rates and exploits temporal redundancy to a small degree. SX is a derivative of MPEG that has been adapted for professional purposes in broadcasting. A more open professional MPEG format – Pro-MPEG – has been standardised but will have to compete in the market place with both DV and SX that have significant commercial forces supporting them.

For a system specifically for broadcasting, the Source Coding and Multiplexing portions are defined via the standards developed jointly by the International Standards Organisation (ISO) and International Electro-technical Committee (IEC). Sub Committee 29 of the 1st Joint Technical Committee of the ISO/IEC - ISO/IEC/JTC1/SC29 - is responsible for the 'Coding of Moving Pictures and Associated Audio'. The 11th Working Group of this committee is known as the Moving Picture Expert Group - or MPEG. A sister group - the Joint Photographic Expert Group, or JPEG, has developed a standard [83] for coding still images. It is vital to understand that MPEG does not specify any transmission channel parameters; this has been left deliberately for other groups, such as the DVB, to deal with (see Chapter 1).

The MPEG has developed appropriate generic Source Coding standards [97, 119, 120] for Video and Audio coding and these have now been accepted and adopted world-wide by broadcasting experts. Beginning with MPEG-1 at low quality transmission at bit rates up to about 1.5 Mbit/s the group also evaluated compatible improvements at rates up to 15 Mbit/s (MPEG-2) which will deal with broadcast quality, at least at the consumer level. A European EUREKA project, VADIS or EU625, was convened to consider the application of MPEG-2 type processes in Europe and had considerable influence on the design of the video and audio coding algorithms chosen in MPEG. The MPEG algorithm is a hybrid of four processes:

- Predictive Coding, which also uses motion estimation and compensation, to exploit temporal redundancy in the moving images,

- Transform Coding which uses the Discrete Cosine Transform (DCT) to exploit Spatial redundancy, and

- Variable Length or Huffman Coding to remove remanent redundancy from the bit stream produced by the first two processes.

- The algorithm also relies upon a buffer store that is used to regulate and smooth the flow of data and whose state of occupancy controls, by means of a feedback path, the coding accuracy. The more full/empty the buffer the coarser/finer the coding becomes in order to reduce/increase the amount of data entering the buffer.

The bit rates quoted above should only be treated as guides. Whereas the applications envisaged at the time of the development of the standards were indeed at these bit rates, CD-ROM and CD- Interactive, for example, time has passed and the provisions of the standards have been used at higher rates. For example, products existed about six years ago that employed the MPEG-1 standard at bit rates from 1 MBit/s up to 15 MBit/s. MPEG-2 can be used from under 1 MBit/s to well beyond the 15 MBit/s nominal limit. The labels are now somewhat historical but, nevertheless, in order to promote interoperability, some outline at least of the MPEG specification must be clear for a number of applications. The specification allows for this through its so-called Profiles and Levels. The latter are simply resolution options whilst the former allow for different applications. Conventional television parameters are accommodated by the Main Profile, whilst others allow for Scalable versions of this Main profile where enhanced definition using hierarchical coding methods can be included in a transmission that can be used by both normal and specially equipped high definition receivers. The High Profile allows for full HDTV. Contrary to what might be expected, this flexibility does not render these latest developments non-standard since the standards are very generic and publish only the rules by which coded bit streams are derived and constructed and are therefore largely independent of practical factors such as bit rates or encoder implementations.

There is also within the MPEG scheme a System Layer that provides for the means to transmit an organised stream of data representing a programme or a set of programmes in multiplex. The data are organised into packets that are 188 bytes long. There are Transport Streams that are responsible for carrying complete programme assemblies that may comprise more than one television programme. These streams also comprise means of synchronising recievers, including Clock References and Presentation Time Stamps (PTS), as well as the data. There are also Programme Streams that carry data from video, audio and data sources which belong together as a service and so need to be maintained as a common entity. Data from an individual coding process, say a video channel or a data channel, is called an Elementary Stream.

The practical exploitation of the wide degree of choice available to a user within the rules of the MPEG standards is beyond the scope of this Chapter. ISO 13818-1, 13818-2, and 13818-3 are the appropriate specifications for the Systems, Video and Audio aspects respectively. A fourth document, ISO 13818-4, gives the rules governing compliance. The Video part is jointly published with the ITU-T whose reference is Recommendation H.262. For MPEG-2 video in particular, there are choices of quality and application expressed through Resolution and through Signal to Noise Ratio - using the Profiles and Levels indicated above - and there is a degree of scalability between the different members of the family. Full ITU-R recommendation 601 quality can be chosen with full allowance for broadcast requirements. HDTV can be also be accommodated as can lower resolutions for appropriate applications.

Other parts of the MPEG 2 specification deal with practical matters. Part 5 defines interfaces and protocols for Digital Storage Media Command and Control (DSM/CC). Part 6 deals with Non- Backwards-Compatible audio coding. Part 9 - Real Time Interface - defines the tolerable variation in the timing and synchronisation parameters of Transport Streams as they pass through systems.

It should be noted that, whilst MPEG has defined a complete solution for audio coding, including multi-channel systems for surround sound and home theatre applications, other proprietary systems have also been adopted. The most notable of these is Dolby Digital AC3 that has been chosen for the US digital terrestrial HDTV system and will be used by some US satellite operators, even those using DVB specifications. In practical terms AC3 and MPEG audio perform similarly well, as would be expected since some of the processing techniques are very similar, but the choice has been made on a wider basis than performance.

5.   BEYOND MPEG2 – MPEG-4, 7 AND 21

MPEG-2 has been established in operational use for more than 5 years. Its performance is optimised at bit rates in excess of about 1 MBit/s. For the future, and in applications where the bit rates are considerably less than this value, a derivative algorithm has been developed that allows, together with other features, novel image manipulation at the decoder. MPEG-4 [121] has much in common with MPEG-2 at the conceptual level but differs by enhancing the system layer to enable the delivery of several image planes within a bit stream that constitutes one service. The video and audio compression algorithms are improved. There is no backward compatibility between MPEG-2 and MPEG-4 that allows the many millions of MPEG-2 decoders already installed in receivers and PCs to be used to decode MPEG-4 streams.

The ability of MPEG-4 to dissect the image, or even synthesise a part of it electronically, into different constituent planes lies at its heart. It will allow non-real cartoon like images to be merged with normal real-world images and, at the decoder, the several planes can be combined, either under direct control of the programme maker, or by the viewer, or both. The artistic potential for new kinds of programmes is immense.

In order for the system to support the new complexities, one of the essential parts of MPEG-4 is a labelling scheme whereby the image planes and contents, especially the objects in each plane, can be identified to machine processes as well as the viewers. This implies some kind of description language and data formats that serve to define the image content. This so-called Metadata is what MPEG-7 emphasises with the intent of providing tools for the system and operators to describe the content and its properties. The concept of metadata as a descriptor has already been identified by programme producers [122, 123, 124] as a necessary future tool that will, for example, allow more machine-based programme content manipulation. This has great value in all parts of the production chain in labelling the video and audio clips so that editing is simplified and can done automatically. It will also assist human operators to search archives for relevant images for insertion into news bulletins, documentaries, etc.

MPEG-4 is a very powerful tool and technologically represents a considerable advance but commercially the investment already made in MPEG-2 systems will mean that further widespread and rapid take-up of MPEG-4 by broadcasters is unlikely. However, for new applications in the Internet environment or future new services that require new receivers anyway the possibilities for MPEG-4 can be considered. Production processes are expensive to change and so one aspect of difficulty with MPEG-4 for broadcasters is the availability and cost of production tools and the training of staff to use them.

MPEG-21 is a much more recent initiative that seeks to assimilate many aspects of content management and unify them. It is ambitious and as it has only just begun in practice and there has been only modest progress. Information can be obtained on all aspects of MPEG from its web address [82]

6.   CONDITIONAL ACCESS

Traditional methods of funding broadcasting have been subscription, including mandatory public levies in the form of licence fees, and advertising. These have the industry served well for many years. The explosive growth in the number of television channels available to the public made possible by digital technology has raised questions concerning future funding mechanisms. In Chapter 1 the regulatory issues were explored and it was established that perhaps there is reason to expect that the mandatory public subscription mechanism may decline in importance as the need to share spectrum becomes less significant. Public service television may come under attack from commercial quarters whose interests will be to meet the mass demand in the market rather than serve all sections of the population.

The means whereby commercial players collect their revenues will also be by subscriptions and advertising but new digital technology has enabled new methods of revenue collection. By encrypting all or some of the transmissions commercial broadcasters can target audiences better and can secure payment of viewing fees. The revenues of the public service broadcaster flow directly from licence fee collection that has the force of law to support it and the services are freely available to anyone with a receiver. Commercial broadcasters must recover their own revenues and so must use PayTV or Conditional Access (CA) technology to do so. Not all of their services need to be protected in this way and usually it is only the premium, high value services such as major sports events etc that require this method. Nevertheless, once in place Conditional

Access technology can do a number of other things that enhance the delivery of material that otherwise would be troublesome in a public free access environment. Obvious examples are programmes that contain offensive or contentious material such as sex, violence, excessively horrific movies etc. Less obvious is targeted advertising to avoid bans in some regions or countries, say for alcohol or tobacco, by disabling the receivers. Satellite broadcasters obviously have this problem to a greater extent than terrestrial or cable broadcasters, where the coverage is more naturally restricted to a small locality. Another less obvious application is to limit the viewing of live sports events in the regions around the venues so that people are not discouraged from attending and so protecting the future of the sport. The use of receiver identification, for example through the post-code of the owner's home, can make this possible.

In digital broadcasting the vastly greater range of services that can be offered needs the additional features that CA technology can bring so that operational flexibility can be augmented by commercial flexibility, an essential element in the viability of commercial broadcasting. Digital streams containing the programme material also contain data that control decryption devices, usually embodied as Smart Cards, in the receivers and messages can individually be sent to all subscribers occasionally to update encryption keys, programme entitlements etc.

The commercial value of some television programmes is very high and the use of encryption naturally leads to attempts at piracy. The dilemma for the broadcaster is how to balance the strength and complexity in the system of encryption with its cost and the risk of piracy. This is not simply a technological issue since there are several methods to protect against sustained attack from commercial pirates not the least of these is covert investigation followed by use of the normal processes of law. However, this is also costly and again a compromise between encryption strength and cost is implied. The commercial value of premium programming needs strong encryption, so much so that the national security agencies in each country take a keen and close interest in the technology and have influence over standards and the exportation of cryptographic elements in equipment. Export licences are needed to ship multiplexing and de-multiplexing products (usually those containing the actual encryption devices) but auxiliary parts of the statutory export licence documentation require that the high computing speeds used in compression encoders and high baud rate modulation equipment also come under scrutiny.

It is clear that publication of open standards for encryption systems is difficult for a number of reasons. Obviously the compromising of security is the concern of the operators and owners of the programmes whilst public bodies such as regulators need to ensure that no unfair advantage is taken by these operators such as the monopolisation of viewers through excessive control over receivers containing decryption devices. However some steps can be taken to mitigate these concerns and, for standards, both the ATSC [110] and DVB [98, 99, 101, 102] have defined some mechanisms whereby some open-ness is made possible in the interests of interoperability.

7.   INTERACTIVE TV

Broadcasting has historically been a part of the "passive" entertainment industry and this has been due to the use of elderly analogue technology that has not changed significantly in decades. Programme material is scheduled and published beforehand so that viewers can plan their viewing to fit conveniently with the rest of their lives. Regular positioning of popular programmes such as soap operas in the schedule is expected by the viewing audience who live and work and watch television by routine and this is respected by the service providers. Less popular programmes find time slots in the less busy periods of the day. One result of this is that the advertising value of these less popular slots is reduced because the expected viewing audience is reduced. These losses are part of the current broadcast model and although social habits among viewers are changing the trend is slow.

This model of programme planning and supply is called the "Push" model because all control is at the transmission end and the viewer's role is simply to consume the programme as and when it is transmitted. The introduction of Video Cassette Recorder (VCR) in the 1970s changed the model somewhat and allowed some viewer choice in viewing time. It is notable that the relatively poor picture quality of the VCR has not deterred the viewer's appetite for a different viewing experience and indeed the VCR has caused the appearance and spread of the video-tape rental market that is now a world-wide business with multi-billion dollar annual revenues. The lesson learned is that when the viewer is given choice he will exercise it and, in the UK at least, it has been proven that the viewer will respond positively to greater viewing choice as expressed in new digital multi-channel services.

Thus, if the viewer reacts positively to more choice, then the availability of even more choice, for example by making the selection of programme more interactive, there should be a greater opportunity for service providers. By moving away from the "Push" model towards an "On- Demand" or "Pull" model, epitomised by the Internet, service providers can offer better satisfaction to viewers. Certain telephone companies in the US and Europe have experimented with Video-On-Demand services where requests for service and its delivery are all achieved through the telephone lines using a technology called Digital Subscriber's Loop (DSL). Although the technology has been proved for some years its adaptation into a viable commercial model has taken time partly because of the significant investment required and partly because of regulatory uncertainty about access rights of the telephone lines to others than the telephone companies themselves. In the UK British Telecom have started to install the necessary technology and services are due to start imminently.

The closest natural competitors to the telephone companies are the cable television companies. Because they own the physical access medium they can install integral return channel technology to allow the viewer to have a communication path back to the cable head and from there to the outside world independently of the telephone company or broadcaster. Indeed in the UK cable television companies have the ability to offer telephone services as well as entertainment and these are often free within the cable coverage area. Regulation then allows trunk access by interconnection with telecommunications operators thus permitting an alternative to the conventional telephone operator. Once the return path exists it can support other things like Internet access and, if bandwidths are made suitably greater than the usual telephone modem, typically up to 56 kBauds but more often less, then access speeds will improve service quality and so give cable operators a commercial advantage.

Interactivity is seen by many as "The Internet". In fact the degree of actual interchange, expressed in terms of volume of data, in a typical Internet session is relatively small. During browsing a fair part of the time is consumed, not in communication but in reading and absorbing the information on the screen. Therefore the capacity needed in the return path is determined by the user's patience, not necessarily by the data flow. The downstream flow rate is more important because the main requirement is for bringing data to the viewer, not taking it away and even a few bytes per second data rate in the return path can be sufficient for many requirements.

As a result of these considerations Interactivity takes on a different significance. What does interactivity actually mean and how can the limited resources available in some media be turned to advantage? This question is very significant to broadcasters because, apart from the usual mail-based communication with their audiences, they have never had such a thing as direct and instantaneous ability to interact with them. Digital technology changes all this by permitting simple return paths to be associated with broadcasts. Whilst the server infrastructure needs putting in place at the broadcaster's end, the viewer needs access to means of return path support. One obvious example is the telephone line and recent systems such as those used by BSkyB and ONdigital in the UK and others in the US have succeeded in generating viewer interest and the support of Internet access.

Whilst interactivity can serve to enhance broadcast services simply by providing Internet access, the real value of interactivity to broadcasters is in the form of linkage to the programmes. With appropriate features in receivers that include significant amounts of storage, a semblance of interaction can be suggested without limiting the programme provider or viewer from extracting benefit. With the aid of software tools that have access to the published schedule of programmes, a hard disc in the receiver can record all programmes likely to have the viewer's interest, as well as those specifically requested, and they can then be offered each time the receiver is used. The viewer can then browse and choose among the stored items as if they were available on demand except that there is no real return link to the sending end at all.

If a physical path is available to the broadcaster then more normal interaction can be supported. The difficulty is arranging how it may be done effectively. Among the major issues are, firstly, finding and planning the spectrum, secondly, dealing with viewer access contention and, thirdly, how much capacity, both instantaneous peak and sustained continuous rate, does the system need in the return direction. Realising that for each terrestrial transmitter or satellite there will possibly be millions of viewers, each requiring occasional access too the head end resources there needs to be sufficient spectrum and a means of dealing with demand in an orderly and efficient manner. Models exist for dealing with these issues and in the DVB there has been recent work to specify return path techniques and protocols for both satellite and terrestrial broadcasting and these will be published when finished. DVB return path specifications have been published for use in CableTV [125], GSM telephone networks [126], DECT telephone networks [127] normal PSDN/ISDN telephone networks [128], Satellite Master Antenna TV (SMATV) [129] and Local Multipoint Distribution System (LMDS) systems [130]. In all cases the use of appropriate high performance modulation and channel coding is vital to the efficiency of the solution. Cable systems have developed their own solutions a notable one being the US Data Over Cable Service Interface Specifications (DOCSIS) system [111].

It remains to be seen whether interactivity will be an operational and commercial success but signs are that, without some capability, broadcasters will lose competitive advantage in a fast moving world where hesitation can lead to failure.

8.   MODULATION AND ERROR CONTROL

This topic is the subject of this book and here it is only necessary to place it in the context of broadcasting applications. Broadcasting is essentially a one way process of transmission where a wide bandwidth is needed to sustain the traditional high quality pictures and sound that the analogue services provide. The main requirement is therefore to provide the maximum bandwidth possible within the spectrum resources and channel quality available. However, as illustrated above, it has recently become necessary for broadcasters to consider return paths of various kinds, perhaps integral with the same infra-structure of the down-stream path, so that new interactive services may be offered in the future. Cable television systems have an advantage here whereas media using spectrum will need to plan and co-ordinate these new facilities in the usual way. There is considerable opportunity in these applications to design novel and efficient modulation and error control codes that provide optimum performance.

The Channel Coding process is necessarily peculiar to each different channel medium and its characteristics have to be accommodated by means of specific coding. In satellite channels employing modulation, which should be seen as a form of coding, the main additive source of transmission impairment is receiver noise, which is virtually random, and interference whose characteristics are not definable accurately. The design of the channel shaping filter bandwidths in relation to the symbol rate is assumed to be optimised. There will be the usual inter-modulation factors to consider due to the non-linearity of the Travelling Wave Tubes at the up-link and the satellite when multi-carrier operation is required. In other media similar peculiarities will be found.

The Nyquist and Shannon criteria define the classical limits of performance in linear channels. Already it is possible to see the performance of modern communication systems approach the Shannon limit. Variations exist where channel impairments are not dominated by classical thermal noise as in the simple Additive White Gaussian Noise (AWGN) model of the text-book. Some of these novel models may include non-linear elements that require computation in both the modulation and de-modulation processes. Intelligent modems will adapt by building channel models in the receiver that can react to changes in the transmission medium over time and can extract reliable information. However, without a return path to allow dialogue between modulator and de-modulator with low delay, systems will not be able to achieve the full degree of closed-loop adaptability. Already the OFDM scheme chosen for the European digital terrestrial broadcasting system has the means to form a dynamic channel state estimation model that is open ended but nevertheless provides significant benefits in performance in the presence of some quite dynamic channel quality variations.

9. THE FUTURE

Digital technology in the form of efficient coding systems together with software tools and powerful processors in receivers form a very potent combination that has already begun to change the face of the broadcasting industry. This will continue and may even gather pace as the benefits become more widely understood by broadcasters and viewers alike. In the UK the rate of take up of new digital services via both satellite and terrestrial media has been exceptionally rapid in relation to comparable consumer technologies in the past. However the continuation of that trend will be necessary for commercial reasons and this remains to be seen.

The introduction of new services has already shown that viewers will respond positively. Whether interactivity, where cable TV system have a practical advantage over the others, will attract viewers still remains to be seen because it depends upon how the viewers perceive the service and how it is sold to them. It also depends upon whether the Internet stance of "pulling" material rather than having it pushed towards them will be accepted. The "couch-potato" model of the viewer is not a hopeful one in this respect.

One future goal shared by governments, regulators and operators alike is the recovery of the spectrum still used to support the current satellite and terrestrial analogue services. These services need to be continued because of the vast investment that the viewers have made in good faith in their analogue receivers. Recovery and re-use of the satellite spectrum used for BSkyB services in the UK is already planned as a result of the strong growth in digital take-up. In the UK alone there are approximately 25 Million television households, each containing an average of 2.5 television receivers. If these are valued at an average of about £250 each the viewers' investment totals to £15.6 Billion and is being compounded by the replacement market for new analogue receivers at a rate of about 4.5 million per annum. If these are also valued at about £250, viewers are currently making a further £1.1 billion annual investment.

If the spectrum is to be released sooner rather than later then these receivers must be replaced as soon as possible by inexpensive digital ones that at least provide the terrestrial free access services that the analogue service provides. This stems the growing population of analogue receivers from which viewers expect to gain value for many years. In the UK the viewer typically changes a TV set every 8-9 years. At the current rate of replacement the 75 million receivers already in the UK market will take almost 17 years to complete, assuming no further analogue sets are sold. Similar calculations can be made for other countries where the same dilemma exists. However, those investors that would be willing to conceive and fund plans to enable this to happen must be able see the return on their investment. The value of such cheap receiver would be in accelerating change and this will be of value to the government and to the commercial players.

Re-deployment of the regained spectrum for other uses than television is likely unless the broadcasters themselves can see how the future for their industry would provide a similar if not better return from the same spectrum than alternatives. Part of the vision for the future must be in the deployment of novel modulation and coding technology that will exploit the spectrum well and allow a wide range of operators offering a wide range of services in addition to traditional entertainment. Broadcasters must become entrepreneurial and seek opportunities to become global service providers rather than remain with a narrow regional focus. This future may well belong more to the commercial players rather than the national broadcasters whose purpose may remain as protectors of national culture. Those well established national broadcasters that have world-recognised brands, such as the BBC, and that are encouraged by their regulators and governments to balance their national and international operations may also be winners in the future.

Notes

1 The theoretical quantisation noise power Nq for a linear quantiser is given by $q^2/12$ where q is the step size of the quantiser, assumed to be constant. Corrections to account for quantiser range occupancy and for the ratio of sampling frequency to power measurement bandwidth need to be applied as does a correction for the subjective weighting of the nominally flat noise spectrum. An 8 bit PCM system gives a theoretical weighted luminance video S/Nq in 5 MHz of about 66 dB. The different headroom requirements of composite and component video signals means that they have different S/Nq values. A good analogue system will have a weighted luminance video S/N of better than about 50 dB, a poor one less than about 30 dB [117].

Chapter 3

MODULATION TECHNIQUES IN DIGITAL TV BROADCASTING

1. PRINCIPLES AND BASIC DEFINITIONS OF DIGITAL MODULATION

DEFINITION 3.1 [132] Modulation is the process of imparting the source information onto a bandpass signal with carrier frequency f_c by the introduction of amplitude, frequency and/or phase perturbations. This bandpass signal is called the modulated signal $S(t)$ and the baseband source signal is called the modulating signal $a(t)$.

In the digital TV (DTV) broadcasting systems the modulator converts the MPEG-2 coded signal into a format suitable for broadcasting over a transmission medium. The conversion is generally performed by taking blocks of $k = log_2 M$ binary digits at a time from the MPEG-2 source sequence $\{a_n\}$ and selecting one of $M = 2^k$ deterministic, finite energy waveforms $\{S_m(t), m = 1, 2, \ldots\}$ for transmission over the channel, as shown in Figure 3.1.

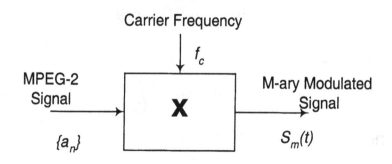

Figure 3.1.   Schematic Diagram of a Modulator

In this case we say that $M - ary$ modulation provides spectral efficiency of:

$$\eta = \frac{log_2 M}{T_0 B} \qquad bit/symb/Hz \qquad (3.1)$$

for statistically independent, equiprobable symbols, where T_0 defines symbol period and B represents bandwidth of the modulated signal.

When the mapping from the digital sequence $\{a_n\}$ to waveforms is performed under the constraint that a waveform transmitted in any time interval depends on one or more previously transmitted waveforms, the modulator is said to have *memory* [131]. Alternatively, when the mapping from the sequence $\{a_n\}$ to the waveform $S_m(t)$ is performed without any constraint on previously transmitted waveforms, the modulator is called *memoryless*.

In addition, the modulation can be classified as being either *linear* or *nonlinear*. Similar to other systems, linearity of a modulation method requires that the principle of superposition applies in the mapping of the digital sequence into successive waveforms. It is apparent that in nonlinear modulation the principle of superposition does not apply.

The most common way to represent a modulated signal is as follows [131]:

$$S(t) = Re\{a(t)e^{j2\pi f_c t}\} \qquad (3.2)$$

Where $Re\{.\}$ denotes the real part of $\{.\}$. The modulating signal $a(t)$ is also called the *complex envelope* of the modulated signal $S(t)$.

In general, any signal is complex-valued and may be expressed as:

$$a(t) = x(t) + jy(t) = A(t)e^{j\theta(t)} \qquad (3.3)$$

Where

$$A(t) = \sqrt{x^2(t) + y^2(t)} \qquad (3.4)$$

and

$$\theta(t) = tan^{-1}\frac{y(t)}{x(t)} \qquad (3.5)$$

Thus

$$S(t) = Re\{A(t)e^{j[2\pi f_c t + \theta(t)]}\} \qquad (3.6)$$

The signal $A(t)$ is called the *envelope* of $S(t)$, and $\theta(t)$ is called the *phase* of $S(t)$.

In digital TV broadcasting the carrier signal is a sinusoid and the characteristics adjusted are amplitude or phase. This is also known as *intermediate* or *radio frequency* (IF or RF) bandpass modulation. The principal reason for employing IF modulation is to transform baseband MPEG-2 encoded signals into signals with more desirable (bandpass) spectra. This allows [134]:

1. Signals to be matched to the characteristics of different delivery media;

2. Signals to be combined using frequency division multiplexing (FDM) and subsequently transmitted using a common physical transmission medium, e.g. cable or satellite transponder;

3. Radio spectrum to be regulated so that interference between different systems is kept to acceptable levels.

In this Chapter we describe major modulation techniques used in DTV.

2. PHASE SHIFT KEYING MODULATION TECHNIQUES

2.1 DEFINITION AND MAJOR PARAMETERS

In $M-ary$ digital phase modulation, also known as *phase shift keying* (PSK), the modulated signal has M different discrete phase values of the carrier:

$$\Theta_m = \frac{2\pi(m-1)}{M}; \qquad m = 1, 2, \ldots M \tag{3.7}$$

Thus, signal waveforms are represented as:

$$S_m(t) = A(t)\cos\left[2\pi f_c t + \frac{2\pi(m-1)}{M}\right] \qquad m = 1, 2, \ldots M, \quad 0 \le t \le T_0 \tag{3.8}$$

where $A(t)$ is the signal pulse shape.

The PSK modulation possess the following properties:

- All signal waveforms have the same energy:

$$\varepsilon = \int_0^T S_m^2(t)dt = \frac{1}{2}\int_0^T A^2(t)dt = \frac{1}{2}\varepsilon_A \qquad m = 1, 2, \ldots M. \tag{3.9}$$

- Signal waveforms may be represented as a linear combination of two orthonormal waveforms, $f_1(t)$ and $f_2(t)$, i.e.

$$S_m(t) = S_{m1}f_1(t) + S_{m2}f_2(t)$$
$$= A(t)\cos 2\pi f_c t \cos \frac{2\pi(m-1)}{M} - A(t)\sin 2\pi f_c t \sin \frac{2\pi(m-1)}{M} \quad (3.10)$$

where the basis functions $f_1(t)$ and $f_2(t)$, and the two-dimensional vectors $S_m = [S_{m_1}, S_{m_2}]$ are given as:

$$f_1(t) = \sqrt{\frac{2}{\varepsilon_A}}A(t)\cos 2\pi f_c;$$

$$f_2(t) = -\sqrt{\frac{2}{\varepsilon_A}}A(t)\sin 2\pi f_c;$$

$$S_m = \left[\sqrt{\frac{\varepsilon_A}{2}}\cos\frac{2\pi(m-1)}{M}, \sqrt{\frac{\varepsilon_A}{2}}\sin\frac{2\pi(m-1)}{M}\right] \quad m = 1, \ldots, M \quad (3.11)$$

Therefore, the space diagram of the MPSK modulation, also known as *signal constellation*, can be represented by M points on the two-dimensional plane. These diagrams for $M = 2, 4$ and 8 are shown in Figure 3.2.

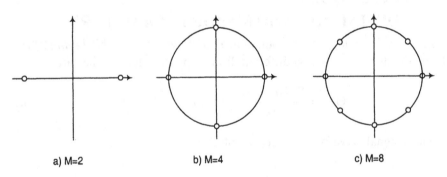

a) M=2   b) M=4   c) M=8

Figure 3.2.   Signal Constellations for MPSK Modulation

Similar to any other modulation format, the performance of the MPSK modulation is defined by the Euclidean distance between the nearest points m and n:

$$d_{min} = S_m - S_n = \sqrt{\varepsilon_A\left[1 - \cos\frac{2\pi(m-1)}{M}\right]} \quad m, n = 1, 2, \ldots M$$
$$(3.12)$$

and for a given signal energy ε_A, the constellation with larger minimum Eucliedan distance will provide better noise immunity and error performance.

Another important problem that needs to be taken into account while designing the physical layer of a broadcasting system with MPSK signals is the problem of mapping $k = \log_2 M$ information bits into the M possible phases of the modulated signal. The solution of this problem depends on the number of factors, such as the type of forward error correction code, channel state, type of noise in the channel, etc. For example, optimum mapping for trellis coded MPSK ($M \geq 8$) signals has been discovered by Ungerboeck in [136], [137] while optimum mapping for uncoded MPSK signals in the poor channel is presented in [138]. However, for uncoded MPSK signals transmitted over the additive white Gaussian noise (AWGN) channel with high *signal-to-noise ratio (SNR)*, the Gray code mapping is the preferred solution.

2.2   BINARY PHASE SHIFT KEYING

Binary phase shift keying (BPSK) is a special class of MPSK modulation in which the modulated signal is allowed to have only $M = 2$ discrete phase values. In principle, any two phasor states can be used to represent the binary symbols but usually antipodal states are chosen with discrete phase values $\Theta_1 = 0^o$ and $\Theta_2 = 180^o$ as shown in Figure 3.2a.

A BPSK modulator with typical baseband and IF waveforms and spectra is shown in Figure 3.3.

It is apparent that BPSK cannot be considered as bandwidth efficient modulation as for data rate $R = 1/T_0 \ bit/sec$ it requires $B = 2/T_0 \ Herts$ bandwidth. Therefore, its application was restricted to early direct-to-home (DTH) satellite broadcasting systems. With the introduction of more sophisticated satellite transponders the use of BPSK signals is gradually reduced.

2.3   QUADRATURE PHASE SHIFT KEYING

A quadrature phase shift keying (QPSK) is one of the most widely used modulation techniques in DTV. It can be represented as a quaternary phase shift keying ($M = 4$) or as a superposition of two BPSK signals with orthogonal carriers. A schematic diagram of QPSK modulator is shown in Figure 3.4.

In this diagram the modulator is effectively two BPSK modulators arranged in phase quadrature, the *inphase (I)* and *quadrature (Q)* channels, each operating at half the bit rate of the overall QPSK system [134].

Every QPSK waveform is represented by $k = \log_2 4 = 2$ binary bits and the spectral efficiency of QPSK is twice that of BPSK. This is because the symbols in each quadrature channel occupy the same spectrum and have half the spectral width of a BPSK signal with the same data rate as the QPSK signal.

In the block diagram of QPSK modulator presented in Figure 3.4 no pulse shaping or filtering is shown. Therefore, the modulated signal at the output of such a modulator is defined as *unfiltered* or *rectangular* phase QPSK [134].

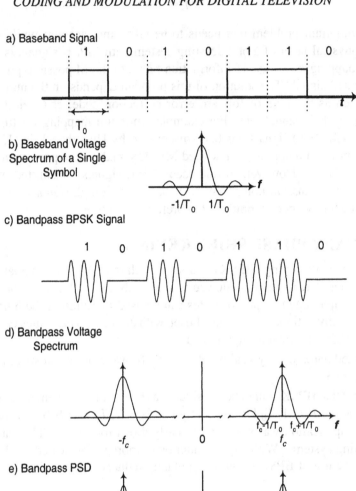

a) Baseband Signal

b) Baseband Voltage
Spectrum of a Single
Symbol

c) Bandpass BPSK Signal

d) Bandpass Voltage
Spectrum

e) Bandpass PSD

Figure 3.3.   Illustration of the BPSK Modulation

The power spectral density of the unfiltered QPSK signal is illustrated in Figure 3.5.

In practical broadcasting systems only filtered QPSK with the appropriate pulse shaping is used. This allows sufficient reduction in the out-of-band spectral components and higher bandwidth efficiency of the overall broadcasting system. Issues related to pulse shaping and baseband filtering will be described in greater detail in Chapter 5.

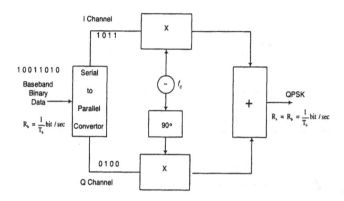

Figure 3.4.   Block Diagram of the QPSK Modulator

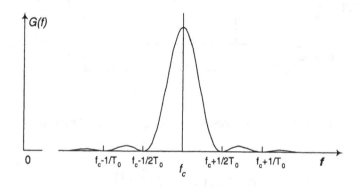

Figure 3.5.   Power Spectral Density of the Unfiltered QPSK Signal

For a given symbol energy, minimum squared Euclidean distance for BPSK signals is $d^2_{BPSK} = 4R^2$ and for QPSK signals $d^2_{QPSK} = 2R^2$, where R is the radius of the circle. As the squared minimum Euclidean distance of the QPSK signal constellation is half of that of the component BPSK constellations, the symbol error ratio of the QPSK is 3 dB worse than that of the BPSK. However, the bit error ratios of the QPSK and BPSK signals are identical. This unique feature of the QPSK signal can be explained by the fact that I and Q components of the QPSK signal are independent and semi-orthogonal and energy per bit, E_b, is the same as illustrated in Figure 3.6.

Thus, the probability of bit error ratio for both the BPSK and QPSK modulation in AWGN channel is given by equation [131]:

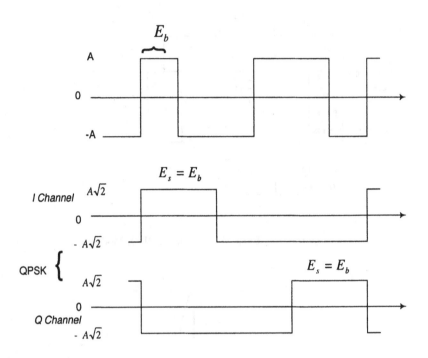

Figure 3.6.   Distribution of Bits Energy in I and Q Channels

$$P_b = \frac{1}{2}\left[1 - erf\left(\frac{E_b}{N_0}\right)^{\frac{1}{2}}\right] \qquad (3.13)$$

Where N_0 is the normalised one-sided noise power spectral density. Alternatively, bit error ratio can be expressed as:

$$P_b = \frac{1}{2}\left[1 - erf\left(T_b B\right)^{\frac{1}{2}}\left(\frac{C}{N}\right)^{\frac{1}{2}}\right] \qquad (3.14)$$

Where the bit period T_b is half of the QPSK symbol period, T_0.

2.4   PHASE SHIFT KEYING WITH INCREASED SPECTRAL EFFICIENCY

Phase shift keying signalling with $M = 8$ different phasor states is recommended by the DVB for Digital Satellite News Gathering (DSNG) with higher bandwidth efficiency [139].

No.	Modulation	Required E_b/N_0 for $P_b = 10^{-4}$	Spectral Efficiency $(bit/s/Hz)$
1	BPSK	8.4 dB	1
2	QPSK	8.4 dB	2
3	8PSK	11.9 dB	3

Table 3.1.   Comparison of BPSK,QPSK and 8PSK Modulation Techniques

As mentioned in Section 3.1, higher values of M lead to the higher spectral efficiency of the modulation. Therefore, spectral efficiency of the 8PSK signalling can be expressed as follows:

$$\eta_{8PSK} = \eta_{QPSK}\frac{\log_2 8}{\log_2 4} = 1.5\eta_{QPSK} \qquad (3.15)$$

However, higher bandwidth efficiency of the 8PSK signalling comes with the degradation of the error performance that needs to be taken into account in order to guarantee the required quality of service. For example, for symbol error ratio $P_s = 10^{-5}$, the difference between QPSK and 8PSK signalling formats in the required signal-to-noise ratio is approximately 4 dB.

The exact expression for bit error ratio of 8PSK is very complicated and not convenient for practical use. However, good approximation can be achieved by applying the formula:

$$P_b = \frac{1}{3}\left[1 - erf\left[\sin\frac{\pi}{8}\left(\frac{3E_b}{N_0}\right)^{\frac{1}{2}}\right]\right] \qquad (3.16)$$

Error performance curves for BPSK, QPSK and 8PSK signalling formats are presented in [131],[134]. For reader's convenience, these results are summarised in Table 3.1.

The signal constellation for 8PSK signalling is shown in Figure 3.2c. Block diagram of the 8PSK modulator can be derived from the equation (3.10) as shown in Figure 3.7.

Recent developments in the ASIC and FPGA design made it possible to implement the 8PSK modulator as a combination of a look-up tables, which generate I and Q components for a given input binary 3-dimensional vector, followed by the corresponding RF circuitry. This allows the reduction in the implementation margin and achievement of data rates of up to 200 $Mbit/s$.

3.   QUADRATURE AMPLITUDE MODULATION
3.1   DEFINITIONS AND MAJOR PARAMETERS

Quadrature amplitude modulation (QAM) represents a class of modulation in which two parameters of the carrier, amplitude and phase, are changed simul-

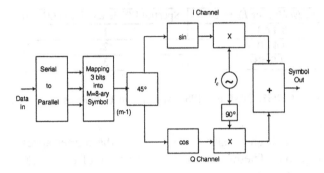

Figure 3.7.   Schematic Diagram of the 8PSK Modulator

taneously with the change of the modulating signal. For a detailed description of QAM signalling format and its derivatives the reader is addressed to an excellent book by Webb and Hanzo [135]. In this Section we describe general properties of QAM and concentrate on its implementation in conjunction with the DVB applications, such as DVB-T, DSNG and DVB-C [139], [144] and [147].

It is apparent that a combination of phase and amplitude shift keying may result in a great number of possible signal constellations. For example, Figure 3.8 illustrates only 2 possible constellations for 16-QAM, but in general, every system designer can develop a constellation that would suit his/her particular requirements.

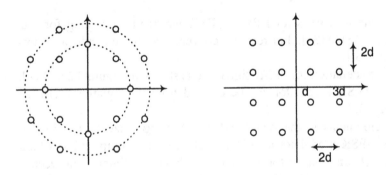

Figure 3.8.   Possible Signal Constellations for 16QAM

When designing a QAM constellation, consideration must be given to [135]:

1. The minimum Euclidean distance amongst phasors, which defines the noise immunity of the developed system;

2. The minimum phase rotation amongst constellation points, which determines the scheme's resilience against synchronisation inperfections and channel phase rotations;

3. The peak-to-average ratio of phasor power, which determines the robustness against non-linear distortions introduced by the broadcasting channel, and specifies the required power back-off to eliminate these distortions;

4. Noise distribution characteristics in the channel.

It has been shown [131],[135],[139],[147] that the square constellation, presented in Figure 3.8a, is optimum for Gaussian channels. Below we will derive some essential characteristics of this square constellation of 16QAM. By applying simple geometric procedure, the minimum phase rotation angle can be estimated as:

$$\Theta_{min} = 26.5^o \tag{3.17}$$

while the minimum Euclidean distance between the phasors is

$$d_{min} = 2d; \tag{3.18}$$

It is known that the energy of a phasor with coordinates (x, y) can be calculated as:

$$E = x^2 + y^2 \tag{3.19}$$

Thus, the average phasor energy for square 16QAM constellation is:

$$E_0 = \frac{1}{16}\left[4(d^2 + d^2) + 8(9d^2 + d^2) + 4(9d^2 + 9d^2)\right] = 10d^2 \tag{3.20}$$

Therefore, the minimum Euclidean distance can be expressed as:

$$d_{min} = 2d = 2\sqrt{\frac{E_0}{10}} \approx 0.63\sqrt{E_0} \tag{3.21}$$

The peak energy of the constellation is:

$$E_{peak} = 9d^2 + 9d^2 = 18d^2 \tag{3.22}$$

Thus, the peak-to-average energy ratio is:

$$r = \frac{E_{peak}}{E_0} = \frac{18d^2}{10d^2} = 1.8 \tag{3.23}$$

It has to be mentioned that the square constellation cannot be constructed for odd-bit values of $M = 2^{2k+1}$, $k = 1, 2, \ldots$ as shown in Figure 3.9.

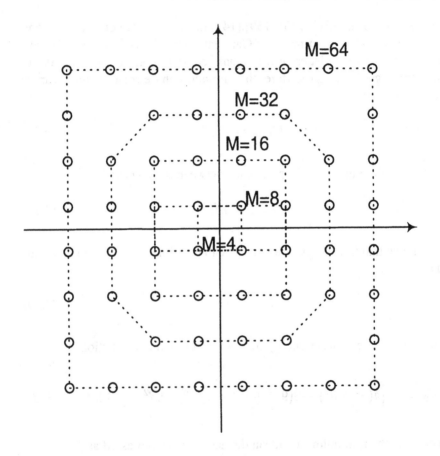

Figure 3.9.   Signal Constellation for Rectangular MQAM

In case of 32QAM this could be considered as an additional benefit since the phasors with the highest energy are not used and peak-to-average ratio is reduced. However, it is not possible to arrange Gray code mapping [141] for such a constellation, therefore more sophisticated combined coding and modulation techniques should be considered [135], [142].

3.2 METHODS OF GENERATING AND DETECTING QAM

Block diagram of a typical MQAM modulator is shown in Figure 3.10.

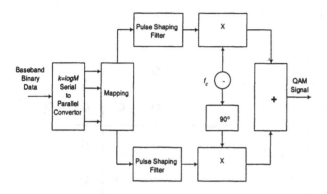

Figure 3.10. Block Diagram of a Typical MQAM Modulator

Similar to MPSK modulator, the input binary data is passed through a $k = \log_2 M - bit$ serial-to-parallel convertor. The parallel data is then passed to a mapper, which creates one-to-one correspondence between the binary input data and the corresponding points. This is followed by pulse shaping filters and up-convertors to the carrier frequency. Usually, pulse-shaped signals are over-sampled in the digital-to analogue (D/A) convertors and the smoothed signal is represented by the corresponding look-up tables. Such a solution demands further filtering after D/A convertion in order to remove aliasing errors caused by the oversampling ratio [135]. However, this filtering is relatively simple to implement.

An alternative method for generating MQAM signals, also known as *superposed QAM*, is illustrated in Figure 3.11 for the case of 16QAM signalling [135],[141].

In this diagram a 16QAM square constellation is constructed using two QPSK modulators. The first QPSK modulator operates at a certain power and the second QPSK modulator operates at one quarter of the power of the first. A block diagram of such a modulator is shown in Figure 3.12.

This method of generating QAM signals is more efficient since it allows the design of MQAM modulators for almost all values of $M = 2^k$ using well developed technology of QPSK modulators. However, the technique has not gained much popularity and only recently started to be explored for the development of hierarchical modulators of the DVB-T standard [144], [145].

In general, the conventional hierarchical modulation is applicable for use with a satellite or terrestrial broadcast service. However, it is more preferebly

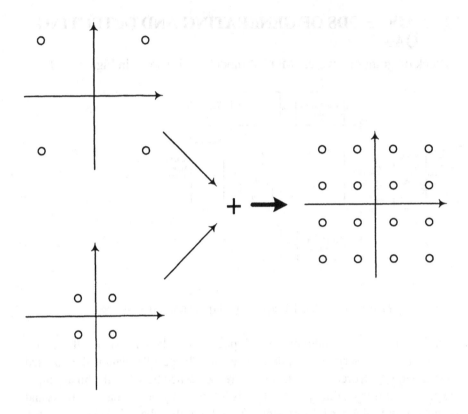

Figure 3.11.   Construction of the 16QAM With Two QPSK Modulators

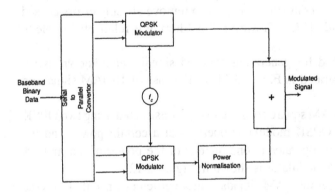

Figure 3.12.   Block Diagram of the Superposed 16QAM Modulator

employed for terrestrial TV broadcast services in which the energy utilisation
and non-linear distortions are not crucial, as compared with satellite broadcast

No.	Modulation	E_b/N_0 loss for $BER = 10^{-2}$	Spectral Efficiency $(bit/s/Hz)$
1	BPSK	0 dB	1
2	QPSK	0 dB	2
3	16QAM	3.98 dB	4
4	64QAM	8.45 dB	6

Table 3.2. Performance Comparison for Uncoded QAM

services. For satellite services other types of hierarchical modulation are proved to be more efficient [146].

Table 3.2 illustrates the loss of performance for different QAM formats in comparison with BPSK modulation. These values define the maximum loss in E_b/N_0 relative to BPSK that is needed to achieve $BER = 10^{-2}$ in an uncoded QAM link. From this table it follows that in order to double data rate in the broadcasting channel (by using 16QAM instead of QPSK) approximately 4 dB of additional E_b/N_0 is required. This energy penalty is increases further increase in data rate is required. For example, difference between 16QAM and 64 QAM (50% increase in data rate) is almost 4.5 dB.

4. VESTIGIAL SIDEBAND MODULATION
4.1 BASIC PRINCIPLES

Vestigial sideband modulation (VSB) can be considered as a compromise between the dual sideband (DSB) and single sideband (SSB) modulation formats [132]. The primary application for VSB modulation is broadcasting of NTSC television signals, which require narrower bandwidth when compared with DSB and, at the same time, a less complex and expensive receiver when compared to SSB. Vestigial sideband modulation can be generated by partial suppression of one of the sidebands of a DSB signal, as illustrated in Figure 3.13.

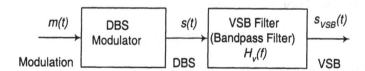

Figure 3.13. Generation of the VSB Signal

In this figure one sideband of the DSB signal is attenuated by using a bandpass vestigial sideband filter that has an asymmetrical frequency response about $\pm f_c$. Thus, the VSB signal can be represented as:

$$s_{VSB}(t) = s(t) * h_v(t) \tag{3.24}$$

where $s(t)$ is a VSB signal, and $h_v(t)$ is the pulse response of the VSB filter. The spectrum of the VSB signal is presented in Figure 3.14

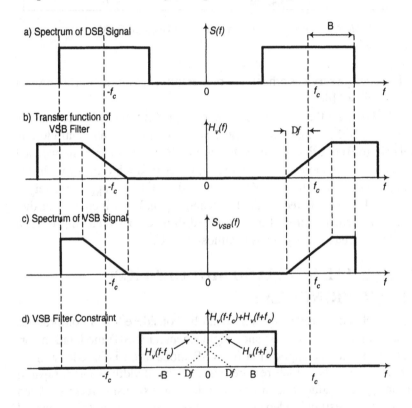

Figure 3.14.   VSB Signal

and can be calculated as:

$$S_{VSB}(f) = S(f)H_v(f) \tag{3.25}$$

where the spectrum of the DSB signal is given as:

$$S(f) = 0.5A_c[M(f - f_c) + M(f + f_c)] \tag{3.26}$$

and $H_v(f)$ is a transfer function of the VSB filter. Thus:

$$S_{VSB} = 0.5A_c[M(f - f_c)H_v(f) + M(f + f_c)H_v(f)] \tag{3.27}$$

A receiver that uses either product or envelope detection can be used to achieve the recovery of an undistorted VSB signal. In both cases the transfer function of the VSB filter must satisfy the constraint (see Figure 3.14d):

$$H_v(f - f_c) + H_v(f + f_c) = C \qquad |f| \leq B \qquad (3.28)$$

where C is a constant and B is the bandwidth of the modulation.

For a detailed proof and explanation of the need for this constraint, the reader is addressed to [132]. Here we note that in NTSC TV broadcast the use of VSB allows the reduction of the required channel bandwidth to $6MHz$. In this application, the frequency response of the vision TV transmitter is flat over the upper sideband out to $4.2MHz$ above the vision carrier frequency and is flat over the lower sideband out to $0.75MHz$ below the carrier frequency [132]. Therefore, the IF filter in the typical NTSC receiver has the VSB filter characteristics shown in Figure 3.14 b, providing that $\Delta f = 0.75MHz$.

4.2   VESTIGIAL SIDEBAND MODULATION FOR ATSC TRANSMISSION SYSTEM

The digital TV standard adopted for the United States allows digital transmission of high quality video and audio signals, in particular high definition television (HDTV), in the same $6MHz$ bandwidth currently used by NTSC. After a series of tests and comparisons, the Grand Alliance selected VSB as the digital transmission subsystem [148],[150],[151],[152]:

- trellis coded 8VSB for terrestrial mode;
- 16VSB for cable mode.

In the terrestrial mode, the 8-VSB signal is a four-level amplitude modulated vestigial sideband signal. Trellis coding transfers the four-levels into eight-level output signal. Rugged system behaviour has been achieved by making the receiver capable of remaining locked below data error threshold. The digital VSB transmission system uses three supplementary signals for synchronisation [150]. A low-level pilot is employed for carrier acquisition, a data segment sync for synchronising the data clock in both frequency and phase, and a data frame sync for data framing and equaliser training.

The VSB and NTSC spectra are shown in Figure 3.15.

Similar to NTSC, the digital VSB spectrum is flat throughout most of the band due to the noise-like attributes of randomised data. This is achieved by the introduction of a pseudo-random scrambling sequence, which flattens the spectrum on average. Two steep transition regions, each $620kHz$ wide, exist at both ends of the transmission band. This is a very efficiently used bandwidth, as there is only 11.5 % excess RF channel bandwidth [150]. The low-level pilot carrier is present at the lower band edge, making the digital VSB signal very power- efficient and reducing its co-channel interference into NTSC [148].

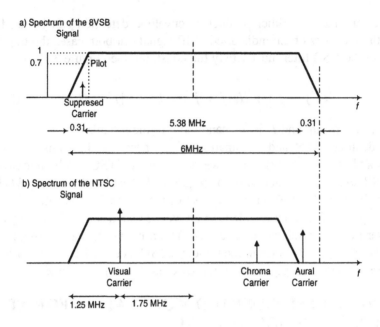

Figure 3.15.   Spectra of the VSB and NTSC Signals

Unlike the NTSC signal, the digital VSB signal is random in nature. There-fore, its average power level is very stable and is therefore used instead of peak power. As reported in [151], 99.9% of the transmitted digital VSB signal peaks are within $6.3dB$ of its average signal power.

The baseband data format for the terrestrial trellis coded 8VSB format is illustrated in Figure 3.16.

Because the MPEG-2 compliant data generated by the Grand Alliance video compression system is segmented into 188 byte packets including a sync byte, data is transmitted in similar segments [148]. Each segment contains 187 data bytes plus 20 parity bytes for forward error correction the the Reed-Solomon code. The last byte replaces the MPEG synchronisation byte, which is re-inserted at the receiver. Over cable, where the signal-to-noise ratio is controlled, a 16-level VSB modulation is used without trellis coding. The increase in the number of levels does not alter the signal's spectrum, but doubles the available data rate, when compared to 8-VSB. More detailed explanation of this diagram will be presented in Chapter 5. Here we emphasise that there are eight discrete data levels and there is a flexibility of selecting any of the required modes:

1. $19.3Mbit/s$ trellis coded 8VSB for terrestrial DTV broadcasting;

2. $38.6Mbit/s$ 16VSB (no trellis) for carrying two $19.3Mbit/s$ HDTV signals in one $6MHz$ channel on cable systems.

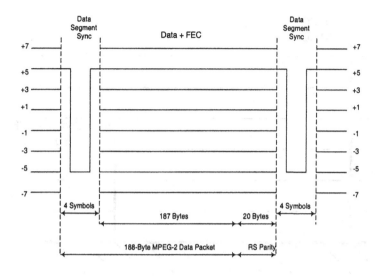

Figure 3.16. Data Segment Format for Trellis-Coded 8-VSB

No.	Parameters	8-VSB	16-VSB
1	Channel Bandwidth, MHz	6.0	6.0
2	Excess Bandwith, %	11.5 d	11.5
3	Symbol Rate, $MSym/sec$	10.762	10.762
4	Bandwidth Efficiency, $Bit/Symbol$	3	4
5	Trellis Coding Rate	2/3	None
6	Reed-Solomon FEC	(207, 187, 10)	(207, 187, 10)
7	Payload Data Rate, $Mbit/sec$	19.3	38.6
8	Peak/Average, dB	6.3	6.4
9	C/N at Threshold, dB	15.0	28.5
10	Phase Noise Threshold, dBc/Hz	-78	-83
11	Impulse Noise Threshold, μsec	193	96

Table 3.3. Characteristics of the VSB Signal

Table 3.3 illustrates the characteristics of both the VSB modes described above.

System robustness is determined by the amount of AWGN that can be handled before data errors occur. Unlike analogue NTSC, which degrades gradually as the signal level decreases, digital transmission system is often referred to as having a "cliff effect" [151]. This means that the digital TV receiver will maintain almost perfect picture right up to a certain threshold level. However, by reducing C/N ratio by $1dB$, all the data will be corrupted by errors, resulting in a frozen picture at the output of the receiver. This is illustrated in Figure 3.17, which compares data error margins for both the NTSC and 8VSB signals.

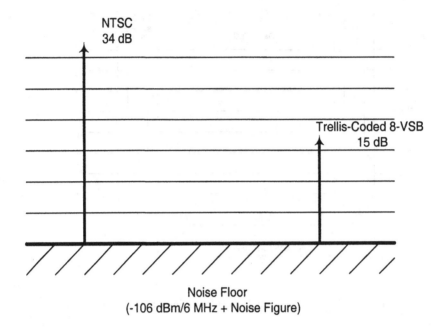

Figure 3.17.   Data Error Margins for NTSC and 8-VSB

It follows from this figure, that analogue NTSC signal requires $34dB$ margin above the noise floor. For trellis-coded 8VSB this margin is reduced by $19dB$ to only $15dB$ above the noise floor. In practice, a $12dB$ "back-off" of the terrestrial digital TV signal will reduce this margin to only $7dB$. However, even such a reduction allows better service availability for the digital 8VSB system, when compared to analogue NTSC.

5.   ORTHOGONAL FREQUENCY DIVISION MULTIPLEXING (OFDM)

5.1   HISTORY OF OFDM

Orthogonal frequency division multiplexing (OFDM) modulation is a special case of a parallel multicarrier transmission, which can be considered as either a modulation technique or a multiplexing technique. In a classical parallel data system, the total signal frequency band is divided into N non-overlapping frequency *subbands* or *subchannels*. In order for the signals to be received independently they must be separated in some sense, i.e. be orthogonal [134]. The traditional way of providing orthogonality in analogue broadcasting applications is to transmit different information signals using different carrier frequencies. This way to separate signals from each other is known as frequency division multiplexing (FDM) and it has been used by communications and broadcasting engineers since the earliest days of radio and telecommuni-cation.

However, frequency division multiplexing leads to inefficient use of available spectrum. An obvious solution to this problem would be to use parallel data and FDM with overlapping subchannels, as illustrated in Figure 3.18. However, the problem of recovery of such a signal represented a major task due to the presence of *intersymbol (ISI)* and *interchannel (ICI)* interference.

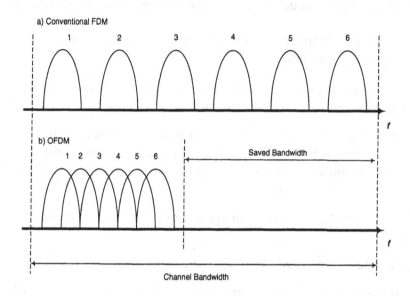

Figure 3.18. Basic Concept of OFDM

The history of OFDM starts in 1966, when a USA patent was filed [156] and Chang published a paper on the synthesis of bandlimited signals for multichan-nel transmission [153]. In this paper he proposes a method for transmitting parallel messages simultaneously through a linear bandlimited channel without ISI and ICI. Following this publication, Saltzberg analysed the performance of Chang's technique and came to the conclusion that " ... the distortions due to crosstalk tend to dominate" [154]. The first practical implementation of OFDM was reported in [155], [161], [162], which described the HF military systems developed by the US army. Similar projects were carried out by the USSR de-fence agency, however, their first open publications appeared after a long delay [165], [166].

One of the major obstacles that delayed the implementation of the OFDM was the need to build a great number of subchannels with all the implications related to the complexity. In 1971 it was proposed to use the Discrete Fourier Transform (DFT) in parallel systems [157]. However, only recently has the silicon industry been able to produce affordable integrated circuits which can be used in a variety of consumer equipment.

In the 1980s OFDM was proposed for high-speed modems, digital mobile communications, high-speed modems over power lines and digital magnetic recording [160]. The use of OFDM has grown dramatically during the last 10 years, as the technique has been implemented for digital audio broadcasting (DAB) [163], digital terrestrial TV [144], and asymmetric digital subscriber lines (ADSL), where it is defined as discrete multitone (DMT) modulation [164].

The main reasons behind such a fast penetration into communications and broadcasting market are [160]:

1. OFDM increases frequency efficiency of the transmission/broadcasting channel, making overall communication/broadcasting system more economical;

2. OFDM provides robustness against narrowband interference, as such interference affects only a small percentage of the subcarrier frequencies. In addition, powerful forward error correction techniques can be used to correct the remaining erroneous carriers;

3. OFDM is particularly attractive for applications in the multipath environment;

4. OFDM makes feasible live TV broadcasting to and from mobile users;

5. OFDM provides a means for the creation of single-frequency TV broadcasting networks.

Obviously, there are also some drawbacks, which restrict the use of OFDM in other applications. These drawbacks will be described in this Section, after detailed explanation of OFDM modulation.

5.2   BASIC PRINCIPLES OF OFDM

The basic principle of OFDM is to divide the available channel bandwidth into a number of subchannels (subcarriers), and to demultiplex the input high-rate information data into these subchannels. By choosing the number of subcarriers to be very large (in the order of a few thousand), the symbol duration will increase correspondingly and the component subchannels will be very narrowband with almost flat fading. This will reduce the relative amount of dispersion in time caused by multipath delay spread and make the equalisation process relatively simple.

In order to obtain a high spectral efficiency, the spectral responses of the component subchannels must be overlapping and orthogonal. In order to eliminate intersymbol interference, a guard interval is introduced in every OFDM symbol. This guard time interval must be longer that the expected delay spread of the signal. In this case multipath components between the two neighbouring symbols cannot interfere with each other. If no symbol is transmitted during the guard time interval, the orthogonality between the subcarriers will be lost, resulting in interchannel interference. In order to eliminate interchannel interference, the OFDM symbol is cyclically extended in the guard time interval by introducing a cyclic prefix [168]. This prefix is a copy of the last part of the OFDM symbol, as shown in Figure 3.19.

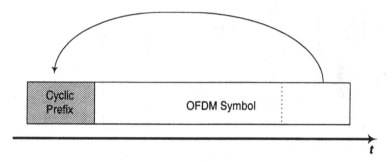

Figure 3.19. Insertion of the Cyclic Prefix

The insertion of the cyclic prefix ensures that the delayed replicas of the transmitted OFDM signal always have an integer number of cycles within the FFT.

Let us consider an OFDM signal that consists of a sum of N_s subcarriers, each one modulated by a QAM. Let also f_c be a carrier frequency of the OFDM symbol and d_i describe the complex QAM symbol. Then, the OFDM symbol starting at $t = t_s$ can be defined as [160]:

$$s(t) = Re\left\{ \sum_{i=-\frac{N_s}{2}}^{\frac{N_s}{2}-1} (d_{(i+\frac{N_s}{2})}) \exp(j2\pi(f_c - \frac{i+0.5}{T})(t-t_s)) \right\} \quad t_s \le t \le t_s + T$$

(3.29)

Where T is the period of the OFDM symbol.

The equivalent complex baseband notation of the OFDM signal is given by the following equation:

$$s(t) = \sum_{i=-\frac{N_s}{2}}^{\frac{N_s}{2}-1} (d_{(i+\frac{N_s}{2})}) \exp(j2\pi\frac{i}{T}(t - t_s)), \quad t_s \le t \le t_s + T \quad (3.30)$$

where in-phase and quadrature parts correspond to the in-phase and quadrature components of the OFDM signal. Therefore, in very general terms, the block diagram of the OFDM modulator can be derived as shown in Figure 3.20.

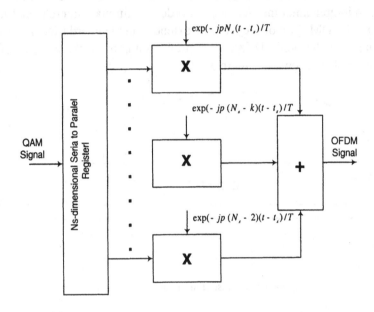

Figure 3.20.   Block Diagram of the OFDM Modulator

In this diagram each subcarrier has exactly an integer number of cycles during the period T, as required to maintain their orthogonality [131]. At the receiver the $j - th$ subcarrier $f_c^j = j/T$ is downconverted and then integrated over the period T [160]:

$$\int_{t_s}^{t_s+T} \exp(-j2\pi\frac{j}{T}(t-t_s)) \times \sum_{i=-\frac{N_s}{2}}^{\frac{N_s}{2}-1} d_{(i+\frac{N_s}{2})} \exp(j2\pi\frac{i}{T}(t-t_s))dt$$

$$= \sum_{i=-\frac{N_s}{2}}^{\frac{N_s}{2}-1} d_{(i+\frac{N_s}{2})} \int_{t_s}^{t_s+T} \exp(-j2\pi\frac{i-j}{T}(t-t_s))dt$$

$$= d_{j+\frac{N_s}{2}}T \tag{3.31}$$

This integration gives the desired output $d_{j+N_s/2}$ subcarrier, while for all other subcarriers the integration result is zero.

Equation (3.30) can be viewed as the inverse Fourier transform (IFT) of the N_s complex QAM symbols. Therefore, the inverse discrete Fourier transform (IDFT) of this signal can be described as:

$$s(n) = \sum_{i=0}^{N_s-1} (d_i \times \exp(j2\pi\frac{i \times n}{N_s})), \tag{3.32}$$

Where n represents sample number. This implies that the generation of the OFDM signals can be implemented by the IDFT, as shown in Figure 3.21, where CE denotes insertion of cyclic extension.

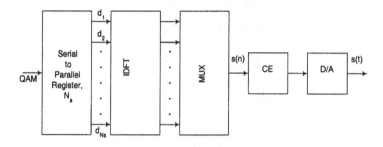

Figure 3.21.   Generation of the OFDM Signals Using IDFT

One of the major issues associated with the generation of the OFDM signals is the problem of the implementation complexity of the IDFT, which grows quadratically with the number of subcarriers, N_s. On the other hand, the complexity of the IFFT is mainly defined by the number of complex multiplications, which grows almost linearly with the number of subcarriers:

$$N_{multiplications} = \frac{N_s}{2} \times \log_2 N_s \tag{3.33}$$

It is apparent that the use of IFFT could significantly reduce the complexity of the desired OFDM modulator. For example [160], in the case of $N_s = 16$ and IFFT based on the radix-2 algorithm [169], provides a reduction in the number of calculations by a factor 8. The beauty of the IFFT is that applying a radix-4 algorithm can reduce its complexity even further. This algorithm can be used to efficiently generate an inverse Fourier transform with a larger size. For more details the reader is referred to [160] and [169].

5.3   SPECTRUM SHAPING OF THE OFDM SIGNALS

As described above, an OFDM symbol is formed by performing an IFFT over N_s subcarriers and adding a cyclic extension. As shown in Figure 3.22, an

OFDM signal is composed from N_s unfiltered QAM subcarriers, which create sharp phase transitions at the symbol boundaries. This results in the out-of-band frequency components, which behave according to a *sinc* function and reduce freqeuncy efficiency of the broadcasting channel [160].

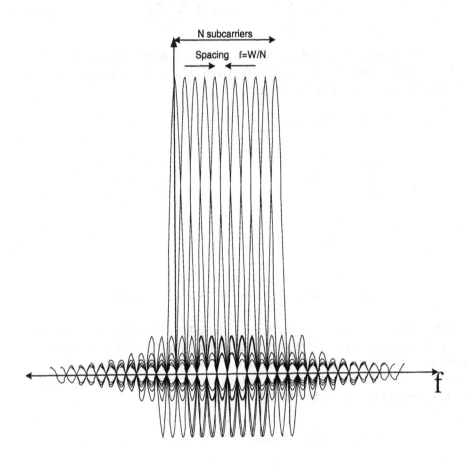

Figure 3.22.   Frequency Response of the Individual Subchannels in an OFDM Symbol

However, special spectral shaping technique, also known as windowing, allows the reduction of the out-of-band spectral components. Windowing operation assumes the multiplication of an OFDM symbol by a window, making spectrum of the output signal a convolution of the spectrum of the window function with a set of impulses at the subcarrier frequencies. In [157], a raised cosine pulse window function is used where the roll-off region also acts as a guard space, as shown in Figure 3.23.

The raised cosine window is defined as:

Figure 3.23.   Pulse Shaping of the OFDM Signal

$$u(t) = 0.5 + 0.5\cos(\pi + \frac{t\pi}{\beta T}) \qquad\qquad 0 \le t \le \beta T_s$$

$$u(t) = 1.0 \qquad\qquad\qquad\qquad\qquad \beta T_s < t < T_s$$

$$u(t) = 0.5 + 0.5\cos(\frac{(t-T)\pi}{\beta T}) \quad T_s \le t \le (1+\beta)T_s \qquad (3.34)$$

Where β is the roll-off factor and the symbol interval T is shorter than the total symbol duration in order to allow the adjacent symbols partially overlap in the roll-off region. The time structure of the OFDM signal is presented in Figure 3.24, while the overall OFDM symbol is defined as:

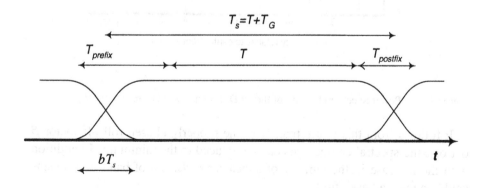

Figure 3.24.   Pulse Shaping of the OFDM Signal

$$s(t) = Re\left\{ u(t - t_s). \right.$$

$$\left. \sum_{i=-\frac{N_s}{2}}^{\frac{N_s}{2}-1} d_{i+\frac{N_s}{2}} \exp(2j\pi(f_c - \frac{i+0.5}{T})(t - t_s - T_{prefix})) \right\} \qquad (3.35)$$

$$t_s \leq t \leq T_s + T_s(1 + \beta)$$

where $t = t_s = kT$ denotes the starting time of the signal. It follows from this equation that larger roll-off factors reduce the out-of-band components and make overall OFDM signal more bandwidth efficient. However, larger roll-off factors lead to a decreased delay spread tolerance, as the roll-off factor of β reduces the effective guard time interval by $\beta \times T_s$. Therefore, a compromise between the desirable delay spread tolerance and maximum acceptable level of out-of-band spectral components needs to be found for every developed system. As an example, Figure 3.25 illustrate the power spectral density of the OFDM signal designed according to the DVB-T specification [144].

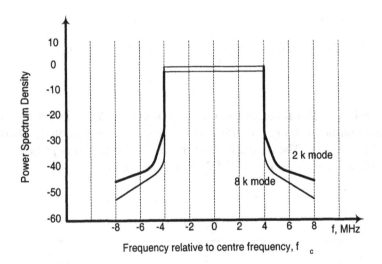

Figure 3.25.   Power Spectrum Density of the OFDM Signal with Guard Interval $T/4$

It follows from this figure that, with the properly chosen roll-off factor β, out-of-band spectral components can be reduced to the minimum. In addition, with the increase in the number of subcarriers N_s, out-of-band components could be reduced even further.

Another approach to reduce the out-of-band spectral components is based on the use of the conventional filtering techniques. In this case a convolution is applied in the time domain and the spectrum of the OFDM symbol is multiplied by the frequency response of the filter [160].

Other types of pulse shaping, such as localised pulses [171] and overlapping [170] also have been investigated in the past, however, their implementation in the DTV systems is limited.

5.4 DEVELOPING OFDM MODEMS

The first step in the development of the OFDM modem is associated with the choice of various parameters of the OFDM signal. To choose the best set of parameters for the desired modem, one needs to find a compromise between a number of opposite and conflicting requirements. Therefore, it is almost impossible to provide general recommendations, which will be applicable to every particular design, as every design case will have different set of initial parameters and requirements.

However, there are three major parameters that need to be specified prior to the design of any OFDM modem:

1. Desired bit rate, R^{des}, bit/sec;

2. Available channel bandwidth, W, Hz;

3. Maximum delay spread, D, sec.

The delay spread defines the guard time interval, which should be chosen between two to four times the root-mean-squared delay spread [144], [160], [172], depending on the type of FEC and QAM used. Once the guard interval is chosen the symbol duration needs to be defined. It is desirable to have symbol duration much larger than the chosen guard interval. However, a smaller duration leads to a narrower spectrum and smaller subcarrier separation. This increases implementation complexity, and makes the overall system more sensible to phase noise and frequency offset [173]. To make things more difficult, a larger symbol duration increases peak-to-average power ratio, making the OFDM signal less tolerant to the non-linear distortions in the amplifier [174]. By taking all these constraints into account, a practical design choice for the symbol duration would be:

$$T \geq 5 \times T_{GUARD} \tag{3.36}$$

where T_{GUARD} is the chosen guard interval.

The next crucial parameter of the OFDM signal, which needs to be determined, is the required number of subcarriers, N_s, which can be calculated as:

$$N_s = \frac{W_{-3dB}}{f_o} = \frac{W_{-3dB}}{\frac{1}{T - T_{GUARD}}} \tag{3.37}$$

where W_{-3dB} is the required $-3dB$ bandwidth of the OFDM signal and f_o represents subcarrier separation.

Alternatively, if the modulation type and error control coding scheme are defined, the required number of subcarriers can be calculated as:

$$N_s = \frac{R^{des}}{R_i} \tag{3.38}$$

where R_i is the bit rate per subcarrier.

EXAMPLE 3.2 The following example illustrates the basics of the OFDM modem design. Let assume that the three major parameters of the desired OFDM modem are specified as follows:
- desired bit rate $R^{des} = 25Mbit/s$;
- maximum delay spread $D = 100\mu sec$;
- bandwidth $W = 8MHz$.

This set of initial parameters is typical for a DVB-T system operating in a flat terrain.

Following the above recommendations, we choose guard time interval as

$$T_{GUARD} = 4 \times D = 400\mu sec$$

and the OFDM symbol duration as

$$T = 5 \times T_{GUARD} = 2000\mu sec$$

Therefore, the subcarrier separation is:

$$f_o = \frac{1}{T - T_{GUARD}} = \frac{1}{2000 - 400} = 625Hz$$

In order to achieve the desired data rate of $R^{des} = 25Mbit/s$, each OFDM symbol must carry k bits, which can be found from the following equation:

$$\frac{k}{T} = R^{des}$$

Therefore:

$$k = T \times R^{des} = 2000 \times 25 = 50,000bit/symb$$

To achieve this number we propose the use of 64QAM modulation in conjunction with $R = 5/6$ forward error correction code. In this case, the number of information bits per subcarrier can be estimated as:

$$k_{subcarrier} = R \log_2 64 = 5bit/symbol/subcarrier$$

Therefore, total number of subcarriers required can be calculated as:

$$N_s = \frac{50000 bit/symb}{5 bit/symb/subcarrier} = 10,000 subcarriers$$

With the chosen subcarrier separation $f_o = 625Hz$, total bandwidth of the dsired OFDM signal will be

$$B = f_o \times N_s = 625 \times 10,000 = 6.25MHz$$

This is less than the bandwidth specified in the requirements. Therefore, we may relax the choice of the FEC and QAM by selecting more powerful FEC with lower code rate (for example $R = 3/4$ or $R = 1/2$) or QAM with smaller number of symbols, i.e. $M = 16$. Alternatively, we can concatenate the chosen channel coding scheme with the RS code which also will provide performance improvement at the expense of additional bandwidth.

6. DEMODULATION OF THE RECEIVED SIGNALS

6.1 DEFINITION OF THE OPTIMUM RECEIVER

Let us consider a modulated signal at the input of the receiver. Following notations adopted in the previous Sections, we assume that the transmitter sends digital TV signals by using $M - ary$ modulation $S_m(t), m = 1, 2, \ldots M$ and each symbol is transmitted within the symbol duration interval, T. We also assume that the channel introduces AWGN only. Therefore, the received signal can be expressed as follows:

$$r(t) = S_m(t) + n(t) \qquad 0 \le t \le T \qquad (3.39)$$

where $n(t)$ denotes AWGN.

The aim of the receiver is to minimise the probability of symbol error by analysing $r(t)$ over the symbol duration interval, T. We will call such a receiver *optimum receiver*. The generic optimum receiver can be represented as consisting of two major components, as shown in Figure 3.26. In this diagram the symbol demodulator converts the received signal $r(t)$ into an $N-$dimensional vector $r = [r_1, r_2, \ldots r_N]$, where N is the dimension of the transmitted signal waveforms. At the next stage, the detector makes a decision which of the M possible waveforms was transmitted. In the rest of this Section we constrain our attention only to the signal demodulators known as *correlation demodulator*. We also assume that no external synchronisation is used, i.e. the receiver recovers synchronisation signal from the received signal $r(t)$.

One of the most important problems in the optimum receiver is to determine the phase of the transmitted carrier. This problem can be solved in one of two

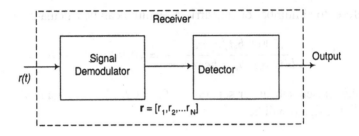

Figure 3.26.   Generic Diagram of the Receiver

possible ways, known as *coherent* and *non-coherent* or *differential* detection. In coherent detectors the phase distortion of the transmitted signal is estimated by a variety of different means. Based on these measurements the receiver can then determine the phase of the transmitted carrier [135]. In non-coherent detection transmitter sends not an absolute phase but differences in phase between the current and previous symbols. The receiver then removes phase distortions introduced by the channel by comparison the previous phase with the phase of the current symbol.

Furthermore, we assume that the filtered received signal is sampled at the right instant and a threshold detector shown in Figure 3.27 makes the decision regarding the transmitted symbol.

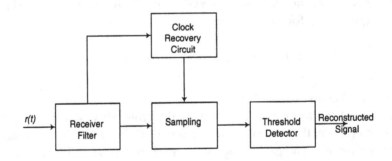

Figure 3.27.   Block Diagram of the Receiver with the Threshold Detector

6.2   CORRELATION RECEIVER

The basic idea behind the correlation receiver is to correlate the received signal $r(t)$ with all legitimate prototype signals $\{f_n(t)\}, n = 1, 2, \ldots N$, using a set of N correlators. At the detection stage, the signal with the highest correlation at instant $t = T$ is assumed to be a transmitted signal.

The prototype signals $\{f_n(t)\}$ can be represented as a series of linearly weighted orthonormal functions, which span the N–dimensional space in such a way that every possible transmitted signal $S_m(t)$, $m = 1, 2, \ldots M$, can be represented as weighted linear combination of basis functions $f_n(t)$, $n-1, 2, \ldots N$.

The block diagram of such a receiver is illustrated in Figure 3.28. In this diagram the received signal $r(t)$ is passed though a bank of N correlators, which operate in parallel and compute the projection of $r(t)$ onto the N basis functions $\{f_n(t)\}$, $n = 1, 2, \ldots N$. Therefore, signal at the input of the $k - th$ correlator can be defined as:

$$r_k = \int_0^T r(t) f_k(t) dt = \int_0^T [S_m(t) + n(t)] f_k(t) dt = s_{mk} + n_k \quad k = 1, 2, \ldots N$$

$$(3.40)$$

where

$$s_{mk} = \int_0^T S_m(t) f_k(t) dt \qquad k = 1, 2, \ldots N \tag{3.41}$$

and

$$n_k = \int_0^T n(t) f_k(t) dt \qquad k = 1, 2, \ldots N \tag{3.42}$$

The overall signal now is represented by the deterministic vector $S_m = [s_{m1}, s_{m2}, \ldots s_{mk}]$, and the component values of this vector depend upon the transmitted signal $S_m(t)$. The noise components n_k are random variables that are determined by the noise in the channel.

EXAMPLE 3.3 Consider an MPSK modulation with the basic pulse shape $A(t)$. We assume that the noise in the channel is a zero-mean AWGN and our aim is to derive the block diagram of the correlation demodulator for MPSK signals.

As described in Section 2.1, all MPSK signals are two-dimensional ($N = 2$, and any signal waveform can be represented as a linear combination of the following basis functions:

$$f_1(t) = \sqrt{\frac{2}{\varepsilon_A}} A(t) \cos 2\pi f_c;$$

$$f_2(t) = -\sqrt{\frac{2}{\varepsilon_A}} A(t) \sin 2\pi f_c; \tag{3.43}$$

where the energy of the pulse is defined as:

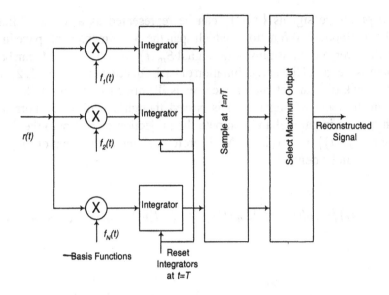

Figure 3.28.   Block Diagram of the Correlation Demodulator

$$\varepsilon_A = \int_0^T A^2(t)dt$$

Thus, the two signals at the input of the threshold detector are:

$$r_1 = \int_0^T r(t)f_1(t)dt$$

$$r_2 = \int_0^T r(t)f_2(t)dt \qquad (3.44)$$

And the generic block diagram of the correlation demodulator for MPSK signals is shown in Figure 3.29.

7.   SYNCHRONISATION SYSTEMS
7.1   INTRODUCTION

There are many levels of synchronisation, such as frame synchronisation, code synchronisation, burst synchronisation, etc. that must be attained in DTV receiver. However, before a receiver can demodulate the received signal, it has to perform at least two major tasks:

1. To find out the beginning of the received symbol and determine its boundaries. This process is known as *clock recovery* or *symbol synchronisation*;

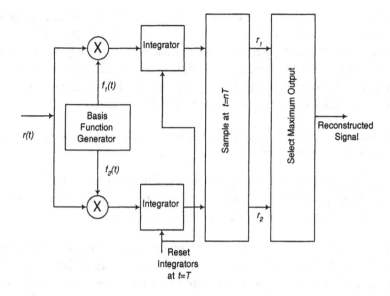

Figure 3.29. Block Diagram of the Correlation Demodulator for MPSK Signals

2. To correct the carrier frequency offset of the received symbol. This process is known as *carrier recovery* or *carrier synchronisation*.

Carrier recovery and clock recovery systems must be considered as an integral part of a demodulation process. Although both systems often operate in a similar manner, it is important to understand crucial difference between them.

Carrier recovery is one of several vital functions of a receiver. It is required in the receivers that use coherent detection that, compared to the non-coherent receivers, provide better error performance for the same channel conditions. There are two basic approaches for dealing with carrier synchronisation at the receiver. One is to multiplex, usually in frequency, a special signal, called a pilot signal, that allows the receiver to extract and, thus, to synchronise its local oscillator to the carrier frequency and phase of the received signal. However, the transmission of a pilot signal will reduce the energy efficiency of the overall system. Thus this technique is not widely used in the commercial digital TV broadcasting systems.

The second approach, which is more prevalent in practice, is to derive the carrier estimate directly from the modulated signal. In particular, such carrier reconstruction implementations are suggested upon examining the gradient of the maximum likelihood function whose solution is the maximum a posterior probability (MAP) estimator of the carrier phase. The MAP approach leads to closed loop implementations with active arm filters which are matched to the signal pulse shape [175]. This approach has the distinct advantage that the total transmitter power is allocated to the transmission of the information-bearing signal. In this book we confine our attention to the second approach.

Carrier synchronisation involves both phase and frequency of the received signal and the carrier recovery system aligns the local oscillator of the receiver with the transmitted carrier frequency [135]. There are two different frequency correction functions that might be needed in a DTV receiver [175]:

1. To aid acquisition of a signal whose initial frequency offset is significant relative to the symbol rate:

2. To provide frequency tracking for differential demodulation of a signal that is not to be phaselocked. It is desired that frequency error during tracking must be reduced to a small fraction of the symbol rate.

The two major categories of phase estimation algorithms are defined as *Decision Directed (DD)* and *Non-Decision Aided (NDA)* algorithms. To achieve the optimum performance, a steady-state phase measurement process requires that all-available information, including data decision and timing information, be utilised. However, DD methods may not behave well when the signal has not yet been acquired. Therefore, independent acquisition of parameters via an NDA algorithm is often necessary [175].

The clock recovery system synchronises the receiver clock with the baseband symbol rate clock introduced by the transmitter. Timing adjustment in this system is achieved by delaying (or shifting phase) of a periodic locally generated signal.

A detailed description of both carrier recovery and symbol synchronisation is presented in [177], [179], [182], [175], [176], [180], [181] and many other books and journal publications. In this Section we present some of the simplest methods that are most widely used in DTV receivers.

7.2   MAXIMUM LIKELIHOOD ESTIMATE CRITERIA FOR CARRIER RECOVERY SYSTEMS

The most general criterion for the development of the optimum carrier recovery technique is the maximum likelihood estimate (MLE) criterion. More strongly, the MLE criterion is the only base we have for a coherent theory of synchronisation. Without the maximum likelihood estimate, carrier recovery techniques can only be seen as "a confusing assortment of different ad hoc techniques with no apparent inter-relation and no a priory clues to relative performance" [175].

Given a signal format and the MLE formulation, it is possible to derive reference-recovery algorithms by formal mathematical operations. The MLE approach is limited to linear, transparent, time-invariant satellite broadcasting channels afflicted only by AWGN. Only M-PSK ($M \geq 2$) modulations are considered and intersymbol interference (after data filtering in the receiver) needs to be negligible. In addition, maximum likelihood estimation requires that the transmitted signal $S_m(t)$ be known at the receiver in most of its basic characteristics: nominal carrier frequency fc, nominal symbol rate $\frac{1}{T}$ and modulation symbol alphabet M. It is the purpose of the MLE to compute estimates φ_e of the carrier phase φ given the knowledge of signal format and the received signal:

$$r(t) = S_m(t) + n(t) = \sqrt{2}A\cos(\omega_c t + \theta_i + \varphi) + n(t) \qquad (3.45)$$

where $S_m(t)$ is the transmitted signal, $\theta_i = \frac{i2\pi}{M}$ $(i = 0, 1, 2 \ldots M - 1)$ is one of the possible phases of the modulated M-PSK signal, $\omega_c = 2\pi f_c$, φ is the initial phase of the carrier signal that needs to be estimated and $n(t)$ is the Additive White Gaussian Noise with spectral density N_0.

Initially, we consider the situation where $S_m(t)$ is completely known except for the parameter φ. The resulting maximum likelihood function, with argument φ_e which can be regarded as a trial estimate of the desired phase, is given by [179]:

$$L(\varphi_e) = \exp\left[\int_0^T [r(t) - S_m(t)]^2 dt\right] \qquad (3.46)$$

The maximum likelihood estimate is the value of φ which minimises the integral in (3.46).This integral expresses the signal space distance between the functions $r(t)$ and $S_m(t)$ defined on the symbol time interval. Alternatively, and equivalently, the integral above can be regarded as a distance between the received noisy signal and a trial local replica signal. The maximum likelihood estimate is that value of φ_e that minimises the distance and makes $S_m(t)$ most nearly like $r(t)$.

The integrand in (3.46) can be broken into the sum of three terms:

$$r^2(t) + S_m^2(t) - 2Re[r(t)S_m(t)] \qquad (3.47)$$

The integral of the first term is the energy of the received signal plus noise and is independent of φ_e.

The integral of the second term is the energy of the local trial signal. It is a function of φ_e and so can influence the maximisation. However, if only carrier frequency and phase are being considered, then the integral of the second term is also a constant and can be ignored [175]. In general, the second term must be taken into account. However, this function is only a weak function of the timing parameter. Therefore, it is common practice to drop this signal from the consideration (neglecting that term could lead to self noise of very small level) [176].

The integral of the third term in (3.47) is often called the correlation between the received signal $r(t)$ and the reference signal $S_m(t)$ so that in this "known-signal" case, the maximum likelihood carrier recovery scheme is a correlator, and φ_e is varied so as to maximise the correlation.

The complex envelope of any linear-modulation signal can be represented in the form [131]:

$$S_m(t) = A \exp(j\varphi) \sum_n \left[a_n h(t - nT - \tau) + j b_n h(t - nT - \tau) \right] \quad (3.48)$$

where $\{a(n), b(n)\}$ represent the complex data sequence and τ is the timing delay.

Correlation and matched filters (or integrate and dump devices) are closely related. For signals represented as (3.48), the correlation can be written as [175], [131]:

$$R = Re\left[\exp(-j\varphi_e) \sum_n c(n) p(n, \tau) \right] \quad (3.49)$$

where

$$c(n) = a(n) + jb(n) \quad (3.50)$$

is the complex value of the $n - th$ symbol and

$$p(n, \tau) = \int_{-\infty}^{\infty} r(t) h(t - nT - \tau) dt \quad (3.51)$$

where $h(.)$ is the channel response function.

It is apparent that integral $p(.)$ is the matched-filter sample, one sample per symbol. This implies that the maximum likelihood estimate φ_e can be extracted from the matched filter sampler. Moreover, it implies that using more than one sample per symbol or a non-matched filter lead to sub-optimum estimates.

Thus, the overall function can be written as follows:

$$
\begin{aligned}
L(\varphi_e) = \\
\exp\left[- \tfrac{1}{2N_0} Re\left[\exp(-j\varphi_e) \ \textstyle\sum_{n=1}^{M} [a(n) + jb(n)] \right]. \right. \\
\left. \textstyle\int_{-\infty}^{\infty} r(t) h(t - nT - \tau) dt \right]
\end{aligned} \quad (3.52)
$$

and after further simplification we will obtain the following equation [176], [177], [181]:

$$L(\varphi_e) = \Lambda = \ln \frac{C}{M} \sum_{i=1}^{M} \exp \frac{1}{2N_0}[I\cos\theta_i + Q\sin\theta_i] \qquad (3.53)$$

where C is a real value constant, I and Q represent the in-phase and quadrophase components of the received signal which can be obtained at the output of the matched filter (or integrate and dump device):

$$I = \int_{-\infty}^{\infty} r(t)\sqrt{2}A\cos(\omega_c t + \varphi_e)dt$$

$$Q = \int_{-\infty}^{\infty} r(t)\sqrt{2}A\sin(\omega_c t + \varphi_e)dt \qquad (3.54)$$

and φ_e is the phase of the local carrier frequency generator.

Further simplification of the maximum likelihood function is desirable and equation that maximises the maximum likelihood function Λ can be written as follows [176]:

$$\frac{d\Lambda}{d\varphi_e} = C\left[th\frac{\sqrt{2}}{2N_0}I\frac{dI}{d\varphi_e} + th\frac{\sqrt{2}}{2N_0}Q\frac{dQ}{d\varphi_e}\right] = 0 \qquad (3.55)$$

This equation can be considered as the general equation for maximum likelihood estimation of the desired carrier phase and it is applicable for any channel condition. However, the closed loop structures which result from this maximum likelihood estimate (3.55) are impractical because of difficulty of implementing the hyperbolic tangent non-linear functions. This equation can be simplified further if we assume the following approximations [178]:

$$thx = sgnx; \quad if \quad x \quad large;$$

$$thx = x - \frac{x^3}{3} \quad if \quad x \quad small \qquad (3.56)$$

Since the input to the non-linearity is a monotonic function of signal-to-noise ratio (SNR), then the approximations (3.55) correspond respectively, to conditions of high and low SNR. As the use of higher order modulation schemes eventually will reduce the signal-to-noise ratio in the channel we assume asymptotically poor channel conditions ($SNR \to 0$) which could be the case for the satellite broadcasting systems using high-order M-PSK systems. In this case the maximum likelihood estimation can be re-written as follows [176]:

$$\frac{d\lambda}{d\varphi_e} = \frac{M^2}{2^{M-1}} \sum_{k=1}^{M} M(-1)^{\frac{k-1}{2}} C_M^k I^{M-k} Q^k = 0 \qquad (3.57)$$

where k is an odd integer number $1 \le k < M$ and C_M^k represent binomial coefficients. It is worth repeating that the above simplification is valid only when $SNR \to 0$ thus carrier recovery technique operating according to (3.57) would be optimal only for low signal-to-noise ratios (obviously the technique will operate in the channels with high SNR however, its performance would be slightly worse than that of the optimum technique).

7.3   BASIC CARRIER RECOVERY SYSTEM

The basic carrier recovery system is shown in Figure 3.30 [182]. It consists of a squaring device, tracking BPF, also known as *phase-locked loop (PLL)*, frequency divider and a delay element.

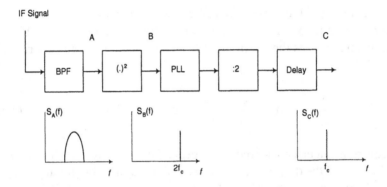

Figure 3.30.   Block Diagram of the Basic Carrier Recovery System

To illustrate the performance of this system we assume the transmission of a BPSK signal in a noiseless channel. The presented analysis could be generalised for a case of any high-order modulation.

Let us assume that the received signal is presented as:

$$S_1(t) = A\cos(2\pi f_c + 0^o)$$
$$S_2(t) = A\cos(2\pi f_c + 180^o) \tag{3.58}$$

where f_c is carrier frequency of the transmitted signal.

As the transmitted signal is assumed to be random and equiprobable, it is apparent that the received spectrum is continuous. The aim of the carrier recovery system is to generate a discrete spectral line at a frequency that is multiple of f_c. This discrete spectral line could be obtained by squaring the received signals:

$$S_1(t)^2 = \left[A\cos(2\pi f_c + 0^o)\right]\left[A\cos(2\pi f_c + 0^o)\right] = \frac{1}{2}C^2[1 + \cos(2\omega_c t)]$$
$$(3.59)$$

and

$$S_2(t)^2 = \left[A\cos(2\pi f_c + 180^o)\right]\left[A\cos(2\pi f_c + 180^o)\right] = \frac{1}{2}C^2[1 + \cos(2\omega_c t)]$$
$$(3.60)$$

From these equations it follows that after squaring, the resulted signal has no dc−component and no phase modulation (both 0° and 180° phase transitions are removed). Furthermore, the signal contains a discrete spectral line at frequency $2f_c$. After PLL, frequency division by a factor of 2 and static delay compensation in the delay element, the recovered carrier frequency is generated.

The squaring and frequency division procedures introduce a 180°-degree phase ambiguity. Therefore, such a carrier recovery circuit could provide the correct replica of the transmitted carrier or its 180°-degree phase shift image. This ambiguity could lead to a reduced error performance or complete loss of signal.

For QPSK signals, quadrupling of the received signal is used instead of a squaring. Quadrupling removes the QPSK phase modulation, however, introduces 90°-degree phase ambiguity.

7.4   ENHANCED CARRIER RECOVERY SYSTEM FOR HIGH-ORDER QPSK

CARRIER RECOVERY SYSTEM FOR QPSK

For QPSK modulation equation (3.55) can be transformed as follows:

$$\frac{d\Lambda}{d\varphi_e} = I^3 Q - I Q^3 = I Q (I^2 - Q^2) \qquad (3.61)$$

and the carrier recovery algorithm can be described as follows:

- if $I^2 - Q^2 = 0$, the phase of the local oscillator can be accepted as the desired estimate of the transmitted carrier;

- if $I^2 - Q^2 = a \neq 0$, the phase of the local oscillator needs to be altered according to the value of a.

One of the possible realisations of this algorithm is shown in Figure 3.31. In this Figure we use the following notations: $Inter.$-integrator, X-multiplier, $(.)^3$-3rd order non-linearity, LPF-low-pass filter, VCO-voltage controlled oscillator. It is important to mention that the developed MLE for QPSK is similar to the technique presented in [175], [178] (although it was derived in a different way).

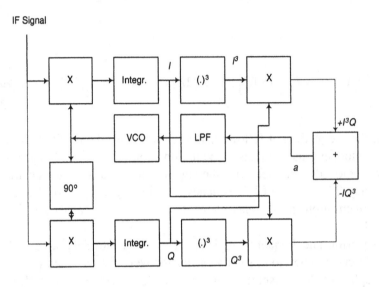

Figure 3.31.   Block Diagram of the QPSK Carrier Recovery System

CARRIER RECOVERY FOR 8-PSK MODULATION

For the case of 8-PSK modulation equation (3.55) can be written as follows:

$$\frac{d\Lambda}{d\varphi_e} = IQ(I^6 - 7I^4Q^2 + 7I^2Q^4 - Q^6) = 0 \qquad (3.62)$$

and the carrier recovery algorithm can be described as follows:
- if $(I^6 + 7I^2Q^4) - (7I^4Q^2 + Q^6) = a = 0$, the phase of the local oscillator can be accepted as the desired estimate of the transmitted carrier;
- if $(I^6 + 7I^2Q^4) - (7I^4Q^2 + Q^6) = a \neq 0$, the phase of the local oscillator needs to be altered according to the value of a.

Figure 3.32 illustrates one of the possible implementations of the carrier recovery system for 8-PSK demodulator (Note: similar to QPSK modulation the above technique is optimum only when $SNR \to 0$).

CARRIER RECOVERY FOR 16-PSK

For the case of 16-PSK modulation the equation for maximum likelihood estimation of carrier phase can be represented as follows:

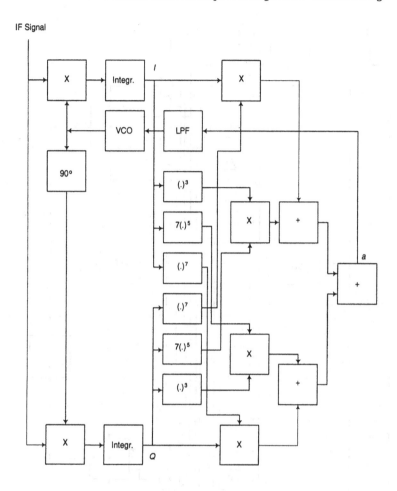

Figure 3.32.   Block Diagram of the 8PSK Carrier Recovery System

Coefficients	b_1	b_2	b_3	b_4	b_5	b_6	b_7
Value	1	273	55	35	35	55	273

Table 3.4.   Coefficients for 16PSK Carrier Recovery System

$$\frac{d\Lambda}{d\varphi_e} = a_1 IQ(I^{14} - 35I^{12}Q^2 + 273I^{10}Q^4 - 55I^8Q^6 \\ +55I^6Q^8 - 273I^4Q^{10} + 35I^2Q^{12} - Q^{14}) = 0 \qquad (3.63)$$

where $a_1 = 2^{-8}$. The block diagram of such a system is presented in Figure 3.33, and the coefficients b_i, $i = 1, 2, \ldots 7$, on this diagram are presented in Table 3.4.

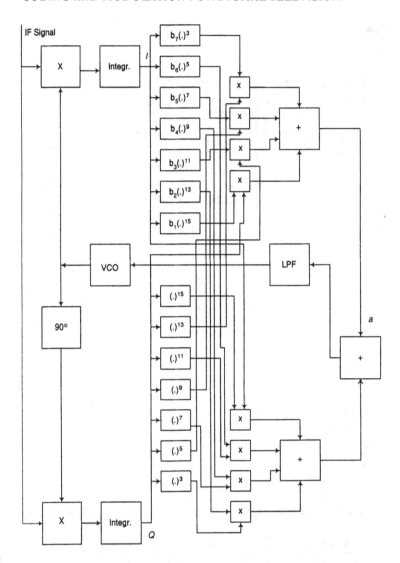

Figure 3.33.   Block Diagram of the 16PSK Carrier Recovery System

IMPLEMENTATION CONSIDERATION

The block diagram of the developed carrier recovery systems (in particular for $M = 16$) may look complicated. However, within the modern digital signal processing architecture there is a possibility of implementing the developed systems using a look-up table stored in ROM-type devices (as shown in Figure 3.34). In this case, the major problem that needs to be solved is the problem of finding the best relation between the received parameters I and Q and the required phase shift of the local carrier oscillator.

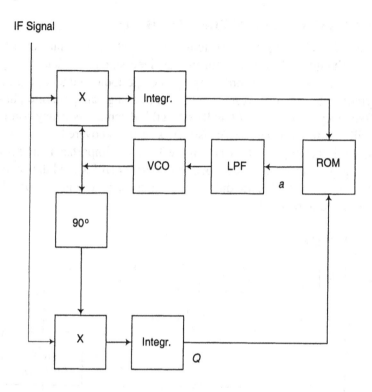

Figure 3.34.   Possible Implementation of the Carrier Recovery System

7.5   CLOCK RECOVERY SYSTEMS

There exists a great number of clock recovery techniques. However, most of these techniques are equivalent as they all make use of the same properties of the received signal. Therefore, all clock recovery systems can be classified as [135], [179]:

1. *Squaring* clock recovery system, also know as *times-two* clock recovery;

2. *Early-late* clock recovery system;

3. *Zero crossing* clock recovery system;

4. *Synchroniser* clock recovery system.

In the rest of this Section we will consider each of these systems. Similar to the carrier recovery case, we will complete our analysis based on the BPSK and show how the presented analysis can be generalised for higher order modulation techniques.

SQUARING CLOCK RECOVERY TECHNIQUE

The squaring clock recovery system is the most fundamental of all clock recovery techniques. In this system the received signal is squared or passed though a non-linear rectifier in order to generate a periodic frequency component at the symbol rate. At the next stage, a PLL or tracking bandpass filter, having a centre frequency at the $f_s = 1/T$ will extract this discrete frequency component which will be used to generate the estimates of the recovered clock. A block diagram of this technique is shown in Figure 3.35. It is important to mention that the squaring clock recovery system performs best for BPSK modulation format, and not so well for higher modulation techniques, where more complicated techniques need to be used.

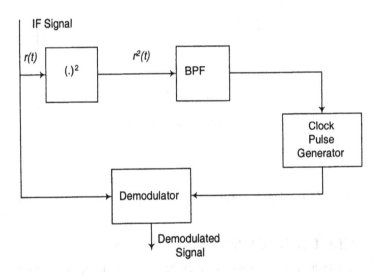

Figure 3.35.   Block Diagram of the Squaring Clock Recovery System

EARLY-LATE CLOCK RECOVERY SYSTEM

The basic principle of the early late clock recovery technique is based on the assumption that the peaks of the received signal are located at the correct sampling instants and that these peaks are symmetrical [183], [135].

The block diagram of the early-late clock recovery technique is presented in Figure 3.36. Similar to the squaring clock recovery technique, the early-late system firstly squares the incoming signal in order to make all peaks of the received waveform positive. At the next stage it analyses two samples of the received signal, chosen in such a way that they are symmetrical relative to the predicted sampling instant. The sample taken prior to the predicted instant is called *the early sample*, while the sample taken after the predicted instant is called *the late sample*. If the predicted sampling instant is chosen correctly then both the early and the late sample will be identical. If the early (late) sample is larger than the late (early) sample, this means that the predicted sampling instant is too late (early).

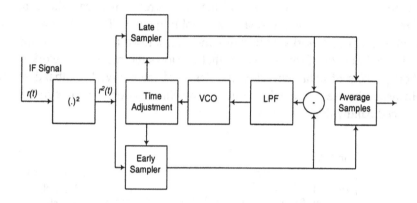

Figure 3.36.   Block Diagram of the Early-Late Clock Recovery System

The early-late clock recovery system works well for modulation techniques in which received waveform is symmetrical around the sampling instant. However, for modulation techniques with partial response raised cosine filtering its performance is not acceptable for DTV requirements.

ZERO CROSSING CLOCK RECOVERY

Similar to early-late clock recovery systems, zero crossing technique also assumes that the received waveform has symmetrical signalling pulses. However, unlike the early-late technique zero crossing clock recovery assumes that the received waveform will pass through zero exactly in the middle between the two sampling instants. The aim of the receiver is to detect the change of the polarity and compare its location with the predicted sampling points. If the change of the polarity occurs exactly in the middle between the two sampling points that the predicted sampling instant is correct. However, if the change of the polarity does not occur in the middle then the reconstructed clock should be altered correspondingly. The block diagram of this technique is similar to the block diagram of the early-late technique with a differentiation unit placed prior to the squaring block.

SYNCHRONISER

This clock recovery system is based on the use of specially dedicated synchronisation sequence, which is transmitted periodically with the broadcasting signal. The block diagram of this technique is shown in Figure 3.37. Such a system is proposed for synchronisation of the terminals in the interactive DTV system, using return satellite channels [184]. In this system the receiver searches for the sequence by performing autocorrelation between the received signal and the replica of the sequence stored in the receiver. In order to increase the accuracy of the method the autocorrelation is calculated at a rate significantly faster than the desired symbol frequency. The technique is simple to implement and works well with almost all known modulation formats. An additional benefit of this system is that it allows the receiver to work with the short packets of data generated as a result of interactive broadband services. However, the use of specially dedicated synchronisation sequence results in the reduced bandwidth efficiency.

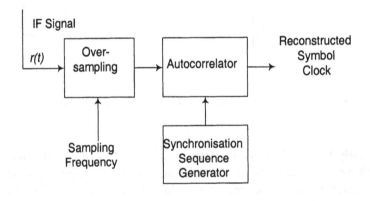

Figure 3.37.   Block Diagram of the Synchroniser

Chapter 4

ERROR CONTROL CODING IN DIGITAL TV BROADCASTING

1. BASIC ELEMENTS OF ERROR CONTROL CODING

The history of error correcting codes can be traced back to the pioneering paper by Shannon [200]. There are essentially two different forms of error correcting methods. The first, which we shall not be dealing with in this book, is that of Automatic Repeat Request (ARQ) schemes, see [203]. These schemes require a return channel (full duplex) system that will acknowledge correct receipt of a piece of data and perhaps resend incorrectly received data. Of course in the main uses of broadcast there is no back channel, the system is half duplex and generally this method is not used. The second is that of Forward Error Correction (FEC), see [203]. In this there is no return channel. The basics are that before sending the data across a channel it is altered so that even if a certain amount of change occurs while sending the receiver will still be able to retrieve the original data that was considered. There are also many hybrid versions of these both singularly and together.

As can be easily seen the FEC scheme is essentially the main contender for the broadcast problem, where the broadcaster wants to omni-directionally send the data to all who can receive. We will now consider the components used in FEC schemes. Later on in the chapter we will explain how all the components are put together in the broadcasting systems we shall be looking at.

2. BLOCK CODES
2.1 INTRODUCTION

One of the first class of block codes to be introduced after the inception of information and coding theory are those of Hamming Codes [196]. Hamming codes are a class of binary single error correcting codes. A major advancement was achieved when Hocquenghem [198] and Bose and Chaudhuri [199] independently discovered a large class of binary multiple error correcting codes. These codes were named BCH codes after the authors of each paper. The cyclic structure of these codes was then discovered by Peterson [197]. The binary nature of these codes was then relaxed by the papers by Gorenstein and Zierler [201]. Just before this discovery an important sub-class of these new non-binary codes was found by Reed and Solomon [202]. Of course just knowing about the construction of these codes is not sufficient for them to be implemented and used in a digital transmission scheme. Before explaining about the construction side in relation to the digital broadcasting we will give the theory necessary to understand and develop block codes. In this book we will only consider linear codes and not non-linear codes as linear codes can be more easily described, and are applicable to the coding schemes used in DVB standards.

We now give a brief introduction to some of the mathematical and algebraic concepts needed to fully understand the coding schemes we will be looking at. For a more thorough explanation see for example [206].

GROUPS

Consider a set G of elements with a defined binary operation on it. Denote this operation $*$, although it may not be multiplication. The operation $*$ on G is a rule that assigns to each pair of elements, a and b a uniquely defined third element, c. We write $c = a * b$. G is closed, that is to say if $a, b \in G$ then $a * b \in G$. The group must satisfy to the following:

1. the binary operation $*$ is associative, i.e. if $a, b \in G$ then $a * b = b * a$,

2. For every element $a \in G$ there exists an element, i such that $a * i = i * a = a$. This element is unique in G.

3. For every element $a \in G$ there exists an element, a^{-1}, known as the inverse of a such that $a * a^{-1} = a^{-1} * a = i$. The a^{-1} is unique for a.

The number of elements in the group is called the *order* of the group.

DEFINITION 4.1 Consider any real number, r. To say we take r mod s for some s we are taking the remainder on dividing r by s.

EXAMPLE 4.2 As in the above definition let $r = 20$ and let $s = 6$. Then

$$\begin{aligned} r \bmod s &= 20 \bmod 6 \\ &= 2 \end{aligned} \tag{4.1}$$

THEOREM 4.3 *For any positive integer m, the set $G = \{0, 1, \ldots, m - 1\}$ is a group under the binary operation $+$ (but taking the result modulo m).*

PROOF. Consider elements $a, b \in G$. Then $a + b = c$ mod m. Now because c is less than m (by the definition of modulo) then this operation is closed. Now integer addition is associative and commutative so $a + b = b + a$ as required. The element 0 is such that $a + 0 = 0 + a = a$, so 0 is the identity element. Now consider the element $m - a$, if we take $a + (m - a)$ then modulo m this is equal to 0, i.e. $a + (m - a) \equiv 0 \bmod m$. So the element $m - a$ is the inverse of a. Therefore the set is a group. $\square$

THEOREM 4.4 *For any prime p, the set $G = \{1, \ldots, p - 1\}$ is a group under the binary operation . (taking the result modulo p).*

PROOF. Consider elements $a, b \in G$. Then $a.b = c \bmod p$. Since p is prime, $a.b$ is not divisible by p and $0 < c < p$, therefore this operation is closed. Now integer multiplication is associative and commutative so it is easy to see that the binary operation . is also associative and commutative. The element 1 is such that $a.1 = 1.a = a$, so 1 is the identity element. Now consider the element a, we need to find an inverse for it. Since p is prime and $a < p$, a and p must be relatively prime. It is well known (for example using the Euclidean algorithm) that there exists two integers b and c such that $b.a + c.p = 1$, with b and c relatively prime. So we get $b.a = -c.p + 1$. Therefore when $b.a$ is divided by p the remainder is 1. There are now two cases:

1. $0 < b < p$, $b \in G$ and it follows that $b.a = a.b = 1$, therefore b is the inverse of a.

2. $b \notin G$, then divide b by p, to get $b = dp + e$. Since b and p are relatively prime, e cannot be zero, and $1 \le e \le p - 1$. Therefore e is in G. So we can get (by combining the above two equations), $e.a = -(c + da)p + 1$, therefore e is the inverse of a.

$\square$

If H is a non-empty subset of G and H is itself a group under the same conditions as G then H is known as a *subgroup* of G.

THEOREM 4.5 *Lagrange: The order of H divides the order of G.*

If $H \ne G$ then H is a *proper subgroup* of G.

EXAMPLE 4.6 Let $G = \{0, 1, 2, 3, 4, 5\}$ be a group under modulo 6 addition. The subset $H = \{0, 2, 4\}$ is a subgroup of G. The group structure can be sen in Table 4.1.

RINGS

We now consider what happens when we add an extra binary operation to a group. Let R be set of elements.

+	0	2	4
0	0	2	4
2	2	4	2
4	4	2	4

Table 4.1.   Addition table for elements in the subgroup

A ring R is a set of that has two binary operations $+$ (generally called addition) and $.$ (generally called multiplication) on it. The elements R must satisfy

1. R is an abelian group under addition, i.e. $a, b \in R \Rightarrow a + b = b + a$.

2. R is closed under multiplication

3. Multiplication is associative, i.e. $a, b, c \in R \Rightarrow a.(b.c) = (a.b).c$.

4. The elements are distributive, i.e. $a, b, c \in R \Rightarrow a.(b + c) = a.b + a.c$ and $(a + b).c = a.c + b.c$.

As the ring R under addition is a group every element in R has an additive identity. The ring R contains a zero element or additive identity and is denoted by 0.

A ring can take on many properties some of which are exemplified below

1. *commutative ring,* R is a ring such that $\forall a, b \in R$, $a.b = b.a$.

2. *ring with identity,* R if the ring R is such that $\forall a \in R$ there exists a multiplicative inverse, 1 and $a.1 = 1.a = a$. The multiplicative identity is unique.

DEFINITION 4.7 An element $a \in R$ that has a multiplicative inverse, which we will denote by a^{-1} is called a *unit.* The set of all units is closed under multiplication. Two elements $a, b \in R$ are called *zero divisors,* if $a.b = 0$ or $b.a = 0$.

A ring R that is a commutative ring with a multiplicative identity that is not equal to the additive identity and has no zero divisors is called an *integral domain.*

DEFINITION 4.8 A ring R that is a commutative ring, has a multiplicative identity that is not equal to the additive identity and every element a has an inverse a^{-1} is called a *field.* We will be visiting these again in the next section.

Let S be a nonempty subset of R. If S is a ring with respect to the same operations on R, then S is called a *subring* of R.

We can see in Figure 4.1 the relationship between rings, fields and all the mentioned classes of rings.

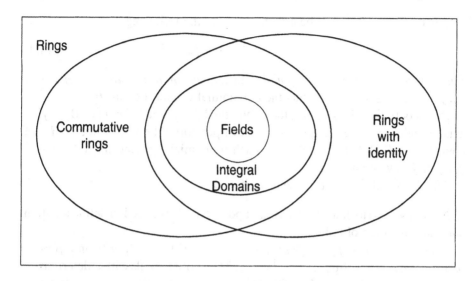

Figure 4.1. Diagram exemplifying the containment of rings

We will now explain some other related topics of rings that will be used further on in the chapter.

DEFINITION 4.9 Let S be a subring of the ring R. If $a.b \in S$ and $b.a \in S$ for all $a \in R$ and all $b \in S$, then S is called an *ideal* of R.

DEFINITION 4.10 Let R be a commutative ring with identity, then the set $\{b.a : b \in R\}$ is called a *principal ideal*. We denote this by $< a >$. This element a of the ring is called the *generator* of the principal ideal.

EXAMPLE 4.11 Let R be the ring of integers, i.e.

$$R = \{-\infty, \dots, -2, -1, 0, 1, 2, \dots, \infty\}. \tag{4.2}$$

Take any element $a \in R$, then the set formed by all the multiples of a, denoted by Ra (or because integer multiplication is commutative we could write aR) is an ideal. Denote this ideal by I, then

$$I = \{-\infty, \dots, -2.a, -1.a, 0, 1.a, 2.a, \dots, \infty\}. \tag{4.3}$$

2.2 POLYNOMIAL RINGS

We want to consider a special ring that has elements that are polynomials. Now, it may seem from first sight that polynomials are complex objects to work with but we will define the addition and multiplication and see that indeed there exists a ring of polynomials.

DEFINITION 4.12 Let R be a ring. Consider the following

$$f(x) = f_0 + f_1 x + f_2 x^2 + \cdots + f_{n-1} x^{n-1} + f_n x^n \qquad (4.4)$$

This object is known as a *polynomial* with *indeterminate* x, with the *coefficients* $f_0, f_1, \ldots, f_n \in R$. We say that the polynomial is over the ring R.

If the coefficient $f_n = 1$ then the polynomial $f(x)$ is called *monic*. By saying x is an indeterminate we mean that it can possibly take values not in R. Of course if $x \notin R$ then the values of each monomial $f_i x^i$ and indeed the sum $f(x)$ may have no meaning.

The *degree* of $f(x)$ is defined as n, if $f_n \geq 1$.

Now, if we are to consider the ring of polynomials we need to define addition and multiplication of polynomials.

Let $f(x) = f_0 + f_1 x + f_2 x^2 + \cdots + f_{n_1-1} x^{n_1-1} + f_{n_1} x^{n_1}$ and $g(x) = g_0 + g_1 x + g_2 x^2 + \cdots + g_{n_2-1} x^{n_2-1} + g_{n_2} x^{n_2}$ be polynomials with indeterminate x and coefficients in a ring R. Without loss of generality let $n_1 \geq n_2$. We define the addition of these two polynomials to be

$$
\begin{aligned}
f(x) + g(x) = & (f_0 + g_0) + (f_1 + g_1)x + \cdots + (f_{n_2} + g_{n_2})x^{n_2} \\
& + f_{n_2+1} x^{n_2+1} + \cdots + f_{n_1} x^{n_1}
\end{aligned} \qquad (4.5)
$$

and define the multiplication of $f(x)$ and $g(x)$ to be

$$
\begin{aligned}
f(x)g(x) = & f_0 g_0 + (f_1 g_0 + f_0 g_1)x + \cdots \\
& + (f_0 g_j + f_1 g_{j-1} + \cdots + f_{j-1} g_1 + f_j g_0)x^j \\
& + \cdots + f_{n_1} g_{n_2} x^{n_1+n_2}
\end{aligned} \qquad (4.6)
$$

The set of polynomials in x over the ring R is denoted $R[x]$. It is left as an exercise to the reader to confirm that using the operations for addition and multiplication defined above that the set $R[x]$ is indeed a ring. $R[x]$ is called the *polynomial ring in x over R*.

THEOREM 4.13 *[205]*

1. If R is a commutative ring then $R[x]$ is a commutative ring

2. If R is an integral domain then $R[x]$ is an integral domain

FINITE FIELDS

A finite field finite field, $F = F_q$ (also denoted GF(q)) is a finite set of q elements with two defined binary (two operands) operations. The operations defined are addition and multiplication. Together with the inverse (that is additive inverse and multiplicative inverse) of each element there exist subtraction and division.

The conditions of finite fields are:

1. F is commutative group under addition. That is $a, b \in F \Rightarrow a + b = b + a$. The identity element with respect to addition is known as the zero element and denoted by 0. There is a unique 0.

2. F is a commutative group under multiplication. That is $a, b \in F \Rightarrow a.b = b.a$. The identity element with respect to multiplication is known as the unit element and denoted by 1. There is a unique 1.

3. The elements of F are associative, i.e. $a, b, c \in F \Rightarrow a + (b + c) = (a + b) + c$, and $a.(b.c) = (a.b).c$. Also the elements are distributive, i.e. $a, b, c \in F \Rightarrow a.(b + c) = a.b + a.c$.

The number of elements in a field is called the order of the field. There exist infinite fields, for example, Q, the field of rational numbers, but we will not be considering them here.

Let F_q^n denote the linear space of all n-tuples over the finite field $F_q =$GF(q). A set V of n-tuples over the finite field F is called a vector space and its elements are called vectors. The field elements are called scalars. For any vectors $\mathbf{u}, \mathbf{v} \in V$, and scalars $a, b \in F$, the following conditions must apply:

1. A vector space V over a field F is a commutative group under addition,

2. The distributive law applies, i.e. $a(\mathbf{u} + \mathbf{v}) = a\mathbf{u} + a\mathbf{v}$ and $(a + b)\mathbf{u} = a\mathbf{u} + b\mathbf{u}$,

3. The associative law applies, i.e. $(ab)\mathbf{u} = a(b\mathbf{u})$,

4. Let 1 be the multiplicative identity of F, then $1\mathbf{u} = \mathbf{u}$,

5. There exist $\mathbf{0} \in V$ and $0 \in F$, such that $\mathbf{u} + \mathbf{0} = \mathbf{0} + \mathbf{u} = \mathbf{u}$, $a\mathbf{0} = \mathbf{0}$ and $0\mathbf{u} = \mathbf{0}$.

The 0 in (5) is sometimes called the origin of the vector space. The addition of two vectors $\mathbf{u} = (u_0, ..., u_{n-1})$ and $\mathbf{v} = (v_0, ..., v_{n-1})$ is defined as $\mathbf{u} + \mathbf{v} = (u_0 + v_0, ..., u_{n-1} + v_{n-1})$, i.e. element-wise addition. Multiplication of a vector $\mathbf{u} = (u_0, ..., u_{n-1})$ and a scalar a is defined as $a\mathbf{u} = (au_0, ..., au_{n-1})$, where each multiplication au_i is performed in F.

These operations of addition of two vectors and multiplication of a vector with a scalar, adhere to the distributive and associative laws.

LINEAR DEPENDENCE AND INDEPENDENCE

Let $v_0, v_1, \ldots, v_{n-1}$, be vectors in a vector space V and $a_0, a_1, ..., a_{n-1}$ be scalars in F. Consider the sum

$$\mathbf{u} = a_0 v_0 + a_1 v_1 + ... + a_{n-1} v_{n-1} \tag{4.7}$$

then **u** is known as a linear combination of the $v_0, v_1, \ldots, v_{n-1}$. A set of n vectors $\{v_0, v_1, \ldots, v_{n-1}\}$ is said to be linear independent if there does not exists a single set of scalars $\{a_0, a_1, \ldots, a_{n-1}\}$, not all $a_i \neq 0$, such that

$$a_0 v_0 + a_1 v_1 + \cdots + a_{k-1} v_{n-1} = \mathbf{0} \qquad (4.8)$$

If there does exist a set of scalars such that this holds then the set of vectors is said to be linear dependent.

In any vector space V, there is at least one set of linearly independent vectors that may generate any vector in V by a linear combination of a (sub)set of the linear independent vectors. This set is called a generating set of V, and if it the smallest such generating set it is known as a basis of V. A generating set is said to span V. The number of vectors in a basis is known as the dimension of the vector space V.

EXAMPLE 4.14 Consider the set of three dimensional vectors over F_2, i.e. $G_1 = \{(1,0,0),(1,0,1),(0,1,1)\}$. G_1 is a generating set of the following vectors:

$$
\begin{array}{c}
(0,0,0) \\
(1,0,0) \\
(1,0,1) \\
(0,1,1) \\
(0,0,1) \\
(1,1,1) \\
(1,1,0) \\
(0,1,0)
\end{array}
\qquad (4.9)
$$

This set of vectors is a vector space. Denote it as V_1.

It is noted here that this is indeed a set of all the vectors of length 3 over F_2. Another generating set is the obvious $G_2 = \{(1,0,0),(0,1,0),(0,0,1)\}$. There is no set of fewer vectors that can generate V_1, so the set G_1 (or indeed G_2) is a basis for the vector space, and the dimension is 3.

SUBSPACES

Consider a vector space V, of dimension k. A subspace , U, of V, is a vector space that is generated by a subset of the vectors in V.

EXAMPLE 4.15 Consider again G_1 as in Example 1.1. We know that this is a basis for V_1. If we take a subset of G_1, for example $G_2 = \{(1,0,0),(1,0,1)\}$, then we can form the vector space, V_2, i.e.

$$(0,0,0)(1,0,0)(1,0,1)(0,0,1) \qquad (4.10)$$

This vector space is a subspace of V_1.

POLYNOMIAL RINGS

We further look at polynomials where the coefficients are from a finite field. Consider the finite field F_q, $q a prime power$ and the polynomial

$$f(x) = f_0 + f_1 x + f_2 x^2 + \cdots + f_{n-1} x^{n-1} + f_n x^n \qquad (4.11)$$

where the $f_i \in F_q$. We denote the set of all such polynomials $F_q[x]$. It is easy to show (in the same fashion as for $R[x]$), that $F_q[x]$ is a ring.

THEOREM 4.16 *[205] Let F_q be a finite field, then every ideal of $F_q[x]$ is a principal ideal.*

DEFINITION 4.17 A polynomial $f(x) = f_0 + f_1 x + f_2 x^2 + \cdots + f_{n-1} x^{n-1} + f_n x^n$, over F_q, is called an *irreducible polynomial* if $f(x)$ is not divisible by any polynomial over F_q other than the constant polynomial (i.e. the polynomial with zero degree), or itself.

EXAMPLE 4.18 Consider the polynomial

$$f(x) = x^3 + x + 1 \qquad (4.12)$$

over F_2. This polynomial is irreducible over F_2 as there is no polynomial of degree greater than zero and less than 3 that is a factor of $f(x)$. However the polynomial

$$f(x) = x^3 + x^2 + x + 1 \qquad (4.13)$$

is not irreducible over F_2, as

$$f(x) = (x + 1)(x^2 + 1) \qquad (4.14)$$

Note all the manipulations are done over F_2.

DEFINITION 4.19 Let $f(x) = f_0 + f_1 x + f_2 x^2 + \cdots + f_{n-1} x^{n-1} + f_n x^n$ be an irreducible polynomial over F_q. If there exists a smallest value $q^m - 1$ (m a positive integer) such that $f(x)$ divides $x^{q^m - 1} - 1$ then $f(x)$ is known as a *primitive polynomial*.

EXAMPLE 4.20 Let $f_1(x) = x^4 + x^3 + 1$. Then $f_1(x)$ is irreducible over F_2. Consider $g(x) = x^{15} - 1 = x^{2^4 - 1} - 1$. Now $f_1(x)$ divides $g(x)$ and no polynomial $x^i - 1$, for $i < 15$ so $f_1(x)$ is a primitive polynomial over F_2.

However, consider $f_2(x) = x^4 + x^3 + x^2 + x + 1$. This, once again is irreducible over F_2. Also $f_2(x)$ divides $g(x)$, but $f_2(x)$ also divides $x^5 - 1$, so $f_2(x)$ is not a primitive polynomial over F_2.

THEOREM 4.21 *Every ideal of the polynomial ring $F_q[x]$ is a principal ideal.*

PROOF. Let I be an ideal of $F_q[x]$. If $I = \{0\}$ then it is the trivial ideal and is obviously $< 0 >$. Suppose that $I \neq \{0\}$, and let $g(x)$ be a minimal degree polynomial in I. Also let $f(x)$ be a non zero polynomial in I. If we divide $g(x)$ by $f(x)$ then we get

$$f(x) = q(x)g(x) + r(x) \qquad (4.15)$$

where the degree of $r(x)$ is less than the degree of $g(x)$. Of course $r(x)$ could equal 0. Since $q(x) \in I$ and $q(x) \in F_q[x]$, then $q(x)g(x) \in I$. Also $f(x) \in I$ and $r(x) = f(x) - q(x)g(x) \in I$. Then since $g(x)$ is of minimal degree $r(x)$ must equal 0. Therefore $f(x) = q(x)g(x)$ and I is the principal ideal $< g(x) >$. $\square$

We now consider another set of polynomial rings formed from the polynomial ring $R[x]$. If we take as an example the ring $F_2[x]$ then we get all the polynomials of the form

$$f(x) = f_0 + f_1 x + f_2 x^2 + \cdots + f_{n-1} x^{n-1} + f_n x^n \qquad (4.16)$$

for any n. If we now take a polynomial, $f(x)$, in $F_2[x]$ and take every polynomial modulo $f(x)$, we get another ring which we denote by $F_2[x]/(f(x))$. The proof that this object is indeed a ring is left to the reader.

It can be seen that as we are reducing modulo a polynomial then the resultant set of polynomials naturally sit in subsets called *equivalence classes*. We take as the representative of each equivalence class the polynomial of least degree in that equivalence class.

EXAMPLE 4.22 Consider $F_2[x]/(x^6 + 1)$. This is a ring. Now define the set

$$I = \{0, 1 + x^2 + x^4, x + x^3 + x^5, 1 + x + x^2 + x^3 + x^4 + x^5\}. \qquad (4.17)$$

It is relatively easy to check that this is indeed an ideal of the ring $F_2[x]/(x^6+1)$. Note here that the ideal $I = < 1 + x^2 + x^4 >$.

We will be lookig again at ideals of polynomial rings when we look at Cyclic Codes.

2.3   FINITE FIELDS

We now consider here more closely a three dimensional vector space V, over F_2. As can be easily seen we have already defined addition on the set of vectors, (that is how we formed the space). However, can we define a multiplication rule on the set of vectors? Consider two elements of the vector space $\mathbf{v}_0 = (v_0^{(0)}, v_1^{(0)}, v_2^{(0)})$ and $\mathbf{v}_1 = (v_0^{(1)}, v_1^{(1)}, v_2^{(1)})$. Is there an easy way to

×	(0,0,0)	(0,0,1)	(0,1,0)	(0,1,1)	(1,0,0)	(1,0,1)	(1,1,0)	(1,1,1)
(0,0,0)	(0,0,0)	(0,0,0)	(0,0,0)	(0,0,0)	(0,0,0)	(0,0,0)	(0,0,0)	(0,0,0)
(0,0,1)	(0,0,0)	(0,0,1)	(0,1,0)	(0,1,1)	(1,0,0)	(1,0,1)	(1,1,0)	(1,1,1)
(0,1,0)	(0,0,0)	(0,1,0)	(1,0,0)	(1,1,0)	(0,1,1)	(0,0,1)	(1,1,1)	(1,0,1)
(0,1,1)	(0,0,0)	(0,1,1)	(1,1,0)	(1,0,1)	(1,1,1)	(1,0,0)	(0,0,1)	(0,1,0)
(1,0,0)	(0,0,0)	(1,0,0)	(0,1,1)	(1,1,1)	(1,1,0)	(0,1,0)	(1,0,1)	(0,0,1)
(1,0,1)	(0,0,0)	(1,0,1)	(0,0,1)	(1,0,0)	(0,1,0)	(1,1,1)	(0,1,1)	(1,1,0)
(1,1,0)	(0,0,0)	(1,1,0)	(1,1,1)	(0,0,1)	(1,0,1)	(0,1,1)	(0,0,1)	(1,0,0)
(1,1,1)	(0,0,0)	(1,1,1)	(1,0,1)	(0,1,0)	(0,0,1)	(1,1,0)	(0,0,0)	(0,1,1)

Table 4.2. Multiplication table for GF(8) in polynomial form

define the multiplication $\mathbf{v}_0\mathbf{v}_1$? First all consider the multiplication table given in Table 4.2.

This table displays all the options for multiplication of each element in V with another element in V. From it we can also get the multiplicative inverses, e.g. the multiplicative inverse of $(1,0,0)$ is $(0,1,0)$ as $(1,0,0).(0,1,0) = (0,0,1)$. However, this appears at first sight to be extremely uninviting and complex. We have however defined our first finite field, the field of 8 elements, where each element is represented by a three dimensional vector. I.e.

$$F_8 = \{(0,0,0),(0,0,1),(0,1,0),(0,1,1),\\ (1,0,0),(1,0,1),(1,1,0),(1,1,1)\}, \quad (4.18)$$

where $(0,0,0)$ is the additional identity and $(0,0,1)$ is the mulitplicative identity. The operations of addition are defined as above i.e. if $\mathbf{v}_0 = (v_0^{(0)}, v_1^{(0)}, v_2^{(0)})$ and $\mathbf{v}_1 = (v_0^{(1)}, v_1^{(1)}, v_2^{(1)})$, $\mathbf{v}_0, \mathbf{v}_1 \in V$, then $\mathbf{v}_0 + \mathbf{v}1 = (v_0^{(0)} + v_0^{(1)}, v_1^{(0)} + v_1^{(1)}, v_2^{(0)} + v_2^{(1)})$, with the addition of the components of the vector being done over F_2. So we could in theory define other finite fields over F_2, and we get a family of finite fields, $F_2^n, n \geq 1$.

Similarly if we start with F_3, then we can get a family $F_3^n, n \geq 1$. Generalising this we get

THEOREM 4.23 *For any prime p and any positive integer n, there exists a finite field F_p^n, of order p^n. No other finite fields exist.*

PROOF. See [206]. □

DEFINITION 4.24 The value p in a field of order p^n is known as the *characteristic* of the field.

THEOREM 4.25 *The characteritic of a field is prime.*

PROOF. This is obvious from Theorem 4.23 □

2.4   POLYNOMIAL REPRESENTATION

Consider now if we rewrite the vectors as polynomials by taking the elements of the vectors as the coefficients of the polynomials. For example $(0, 1, 1)$ is $\alpha + 1$, $(1, 1, 0)$ is $\alpha^2 + 1$, etc. Then we can consider the finite field F_8 as described above as

$$F_8 = \{0, 1, \alpha, \alpha^2, \alpha + 1, \alpha^2 + \alpha, \alpha^2 + \alpha + 1, \alpha^2 + 1\} \qquad (4.19)$$

A similar table to Table 1 can be formed using this terminology but this is not really the best way to approach the problem, it is appears too haphazard, with no obvious structure. In fact there is a lot of structure and it is from this polynomial approach that we can retrieve this structure.

Consider addition of two elements of F_8 as being simply addition of two polynomials. It is easily seen that the addition of any two elements is still an element of F_8. Before considering multiplication we must define and explain the following.

REDUCING POLYNOMIALS MODULO ANOTHER POLYNOMIAL

Let f and g be polynomials with coefficients over any field in the indeterminate x of degree $n - 1$. Formally let

$$f(x) = f_0 + f_1 x + ... + f_{n-1} x^{n-1} \qquad (4.20)$$

$$g(x) = g_0 + g_1 x + ... + g_{n-1} x^{n-1} \qquad (4.21)$$

and consider another polynomial $p(x) = p_0 + p_1 x + \ldots + p_{n-1} x^{n-1} + x^n$. If we form the product of f and g we result in a polynomial of degree $2n - 2$ as follows:

$$f(x)g(x) = f_0 g_0 + (f_1 g_0 + f_0 g_1)x + (f_2 g_0 + f_1 g_1 + f_0 g_2)x^2 + \cdots + f_{n-1} g_{n-1} x^{2n-2}. \qquad (4.22)$$

Now if we want a reduce this polynomial by $p(x)$, also known as taking the modulus of $f(x)g(x)$ by $p(x)$ or $f(x)g(x) \bmod p(x)$, we need to substitute higher degrees of $p(x)$ into $f(x)g(x)$.

First of all we have

$$x^n = p_{n-1} x^{n-1} + p_{n-2} x^{n-2} + \cdots + p_1 x + p0 \qquad (4.23)$$

and further

$$
\begin{aligned}
x^{n+1} &= p_{n-1} x^n + p_{n-2} x^{n-1} + \cdots + p_1 x^2 + p_0 x \\
x^{n+2} &= p_{n-1} x^{n+1} + p_{n-2} x^n + \cdots + p_1 x^3 + p_0 x^2 \\
&\;\;\vdots \\
x^{2n-2} &= p_{n-1} x^{2n-3} + p_{n-2} x^{2n-4} + \cdots + p_1 x^{n+3} + p_0 x^{n+2}
\end{aligned}
\qquad (4.24)
$$

Then we simply substitute in the values for $x^{2n+2}, x^{2n+1}, \ldots, x^{n+1}, x^n$ in that order into $f(x)g(x)$. It is noted here that the resulting polynomial is of degree $n - 1$.

To keep this idea general the equations get very messy and so for clarity we will give an example.

EXAMPLE 4.26 Let $f(x) = x^3 + 2x^2 + 6x + 1$, $g(x) = x^2 + 4x + 1$ and $p(x) = x^3 + 5x + 1$. The coefficients are all integers. What is $f(x)g(x) \bmod p(x)$?

First $f(x)g(x) = x^5 5 + 6x^4 + 15x^3 + 27x^2 + 10x + 1$. We can reduce this modulo $p(x)$ by substituting

$$
\begin{aligned}
x^5 &= -5x^3 - x^2 \\
x^4 &= -5x^2 - x \\
x^3 &= -5x - 1
\end{aligned}
\tag{4.25}
$$

So

$$
f(x)g(x) \bmod p(x) = -4x^2 - 40x - 39. \tag{4.26}
$$

EXAMPLE 4.27 Let us consider an example of polynomials over a finite field. Let $f(x) = x^2 + x + 1$, $g(x) = x^2 + 1$ and $p(x) = x^3 + x + 1$. The coefficients are binary bits i.e. zero or one. For these what is $f(x)g(x) \bmod p(x)$?

First $f(x)g(x) = x^4 + 1$. We now reduce this modulo $p(x)$ using

$$
\begin{aligned}
x^4 &= x^2 + 1 \\
x^3 &= x + 1
\end{aligned}
\tag{4.27}
$$

to give $f(x)g(x) \bmod p(x) = x^2$.

So using this knowledge we can define a multiplication rule for use in a finite field, when the polynomial representation is used. If we once again consider F_8, defined above as (once again with coefficients over F_2).

$$
F_8 = \{0, 1, \alpha, \alpha^2, \alpha + 1, \alpha^2 + \alpha, \alpha^2 + \alpha + 1, \alpha^2 + 1\} \tag{4.28}
$$

then if we have $p(x) = \alpha^3 + \alpha + 1$ and if all multiplication is done by taking the result modulo $p(x)$ we see that the rules of a finite field hold.

EXAMPLE 4.28 Consider two elements α^2 and $\alpha + 1$. Multiplying these two gives $\alpha^3 + 1$, but this is not an element in F_8 so we reduce it by $p(x) = \alpha^3 + \alpha + 1$. This gives α, as expected.

Similarly we can obtain the whole multiplication table as Table 4.3.

Let us generalise things a little now. Consider a field F and let $a \in F$. Then there exists a smallest value n such that $a^n = 1$ (multiplication done in the field F. The values n is called the *order of the element* a in the field F.

×	0	1	α	$\alpha+1$	α^2	α^2+1	$\alpha^2+\alpha$	$\alpha^2+\alpha+1$
0	0	0	0	0	0	0	0	0
1	0	1	α	$\alpha+1$	α^2	α^2+1	$\alpha^2+\alpha$	$\alpha^2+\alpha+1$
α	0	α	α^2	$\alpha^2+\alpha$	$\alpha+1$	1	$\alpha^2+\alpha+1$	α^2+1
$\alpha+1$	0	$\alpha+1$	$\alpha^2+\alpha$	$\alpha^2+\alpha+1$	$\alpha^2+\alpha+1$	α^2	1	α
α^2	0	α^2	$\alpha+1$	$\alpha^2+\alpha+1$	$\alpha^2+\alpha$	α	α^2+1	1
α^2+1	0	α^2+1	1	α^2	α	$\alpha^2+\alpha+1$	$\alpha+1$	$\alpha^2+\alpha$
$\alpha^2+\alpha$	0	$\alpha^2+\alpha$	$\alpha^2+\alpha+1$	1	α^2+1	$\alpha+1$	α	α^2
$\alpha^2+\alpha+1$	0	$\alpha^2+\alpha+1$	α^2+1	α	1	$\alpha^2+\alpha$	α^2	$\alpha+1$

Table 4.3.   Multiplication table for GF(8) in polynomial form

THEOREM 4.29 *Let F be a field of q elements. Then*

1. *Let $a \in F$, $a \neq 0$, then $a^{q-1} = 1$.*

2. *For any element $a \in F$, the order of a divides $q - 1$.*

PROOF. See [205]. □

DEFINITION 4.30 Let F be a field of order $q - 1$. If $a \in F$ has order $q - 1$, then a is called a *primitive element*.

THEOREM 4.31 *Every finite field contains at least one primitive element*

PROOF. See [206]. □

EXAMPLE 4.32 Consider the finite field of 7 elements, which we write as $F_7 = \{0, 1, 2, 3, 4, 5, 6\}$. Addition and multiplication are defined modulo 7. The characteristic of this field is 7.

Now, the element 2 has order 4, as $2^4 = 1$ mod 7. The element 3 has order 6, as $2^6 = 1$, so 3 is a primitive element. The powers of 3 generate all the elements of F_7.

It is can be seen that the primitive polynomial introduced in Definition 4.19 has a root as the primitive element introduced in Definition 4.30.

We need now to consider the relationship between the finite fields F_p, and F_q, where $q = p^n$ for some n. In the example above we have seen a comprehensive explanation for the finite fields F_2 (which we denote the *ground field* and F_8 (which we denote the *extension field*). We will now try to generalise this a little which will hopefully exemplify the subject. let us show an example where we see that the roots of an irreducible polynomial over a ground field lie in the extension field.

EXAMPLE 4.33 Consider the polynomial $f(x) = x^3 + x^2 + 1$ over F_2. This polynomial is irreducible over F_2. It does however, have roots in the extension field F_8 (generated by $p(x) = x^3 + x + 1$, as above). Here we take α to be a primitive element of the field (or alternatively a root of the primtive polynomial $p(x)$). These roots are α^3, α^5 and α^6. We can thus see the field F_8 written in another way,

Power Representation	Polynomial Representation	Vector Representation
0	0	000
1	1	001
α	α	010
α^2	α^2	100
α^3	$\alpha + 1$	011
α^4	$\alpha^2 + \alpha$	110
α^5	$\alpha^2 + \alpha + 1$	111
α^6	$\alpha^2 + 1$	101

Table 4.4. Table of F_8 generated by $p(x) = x^3 + x + 1$

We have previously seen the middle and last columns representing the field, however we can simply use the first column instead. Each column has its own advantage for use. Obviously the first column for multiplication and the second (or third column) for addition of elements. But as is easily seen they all represent the same object.

THEOREM 4.34 *Let $f(x)$ be an irreducible polynomial of degree n over the finite field F_p and $\beta \in F_q$, $q = p^m$. If β is a root of $f(x)$ then β^p, β^{p^2}, ..., $\beta^{p^n - 1}$ are all the roots of $f(x)$.*

DEFINITION 4.35 *Let β be an element in an extension field F_{p^n}. A polynomial $m(x)$ of smallest degree with coefficients in the ground field F_p is called the minimal polynomial of β if $m(\beta) = 0$. This polynomial is irreducible in the ground field.*

It is noted that we could start with ground field as $F_{p^{n_1}}$ and consider the extension field $F_{p^{n_1 n_2}}$. The above definitions and theorems hold for this more general case.

EXAMPLE 4.36 Consider the (extension) field F_8, generated by the (primitive) polynomial $p(x) = x^3 + x + 1$. Also consider the elements α, α^2 and α^4. It is easy to see that $f(x) = x^3 + x + 1$ is a minimal polynomial of α. Note here that $f(x)$ is irreducible over F_2 as noted in Definition 4.35.

3. LINEAR BLOCK CODES

Let us assume in this section that we have a stream of binary digits to send across a channel. If we just modulate the bits then there is no error correction possibility. We need to consider how we can somehow protect what we send. The easiest way to think about this is to add redundancy. Let us consider generally that we have a stream of k information bits, we encode (consider this as a simple conversion) these to n bits, where $n > k$. As an example take the example of a repetition code, C, with $n = 3$ and $k = 1$. If we look at all the information bits that can be encoded then, of course, we have 0 and 1. Let the encoding procedure be as in Table 4.5.

Information Bit	Encoded Word
0	000
1	111

Table 4.5.   Encoding procedure for our $n = 3$, $k = 1$ code

Define the Hamming distance of two words, $\mathbf{x} = (x_0, x_1, \ldots, x_{n-1})$ and $\mathbf{y} = (y_0, y_1, \ldots, y_{n-1})$ to be number of places where $\mathbf{x}$ and $\mathbf{y}$ differ, and denote by $d(\mathbf{x}, \mathbf{y})$. So in the example here we have two code words, and $d(000, 111) = 3$. Now it is easily seen that the power of the scheme we are using here is that if 1 error occurs in a transmitted word, we can still recover the original information bit. Formally put, the repetition code we are considering here has an error correcting capability of 1 (which we define as t). This code, C, is linear because if we take the sum of the two codewords then we get a codeword.

Now we can define the minimum distance of a code, C.

$$d_{min}(C) = \min\{d(\mathbf{x}, \mathbf{y}) : \mathbf{x}, \mathbf{y} \in C, \mathbf{x} \neq \mathbf{y}\} \tag{4.29}$$

So for this simple repetition code $d_{min} = 3$. It is easily seen that

$$t = \left\lfloor \frac{d-1}{2} \right\rfloor \tag{4.30}$$

We can further define the weight of a word, $w(\mathbf{x})$, to be the number of non zero elements of the word $\mathbf{x}$. If we have a code C with parameters n, k and d_{min}, then we say that C is an (n, k, d_{min}) code. Usually d_{min} is written simply as d as in the context the meaning is understood. So our repetition code is a $(3, 1, 3)$ code.

THEOREM 4.37 *The minimum distance of a linear code is equal to the minimum weight of its non-zero codewords.*

PROOF.

$$
\begin{aligned}
d_{min} &= \min\{d(\mathbf{x}, \mathbf{y}) : \mathbf{x}, \mathbf{y} \in C, \mathbf{x} \neq \mathbf{y}\} \\
&= \min\{w(\mathbf{x} + \mathbf{y}) : \mathbf{x}, \mathbf{y} \in C, \mathbf{x} \neq \mathbf{y}\} \\
&= \min\{w(\mathbf{v}) : \mathbf{v} \in C, \mathbf{v} \neq \mathbf{0}\} \\
&= w_{min}
\end{aligned}
\tag{4.31}
$$

□

Define now, the rate, R, of a code to be the ratio of k/n, then we see that this repetition code has rate $1/3$. The code we will consider here is one of the simplest error correcting codes and serves purely as an example. We must now move to more complex codes, so that we can send more than just a 0 or a 1 across a channel. But how do we do this easily? Let us consider the repetition code and try to generalise the situation further.

As we are only considering linear block codes it is easy to see that we can form a generating set for the code. This we will show is fundamental to understanding the properties of many linear codes. Take for example the $(3, 1, 3)$ repetition code we have been using. A generating set for it is consists of only one element, i.e $(1, 1, 1)$. If we write this as G for generator then if u is the information bit we can encode in the following fashion (where **c** is the encoded word):

$$\begin{aligned} \mathbf{c} &= uG \\ &= u.(1, 1, 1) \end{aligned} \qquad (4.32)$$

where the multiplication is done as a scalar multiplied by a vector. So as we saw in Table 3, if $u = 0$ we get $\mathbf{c} = (0, 0, 0)$ and if $u = 1$ then $\mathbf{c} = (1, 1, 1)$, as expected. Usually the vector notation is simply represented without commas and as we shall see later this shall be extended to matrices in a similar way. So equivalently we get

$$\begin{aligned} \mathbf{c} &= u.G \\ &= u.(111) \end{aligned} \qquad (4.33)$$

Let us now consider that we can have a matrix G, of dimension $(n \times k)$ such that we want to encode a vector **u** of length k to a codeword **c** of length n. For the explanation here to be meaningful we must take the rows of G (i.e. the generating vectors) to be linearly independent, otherwise we can remove some of them. So we are generalising the situation above to take the information bit to be an information word and the generating set of the code is indeed now more than one vector. We have now (notice that **u** is now a vector not a single element as in the $(3, 1, 3)$ repetition code):

$$\mathbf{c} = \mathbf{u}.G \qquad (4.34)$$

Define G to be the generator matrix of the (n, k, d) linear code. We know here the values of n and k but we do not know the d. This is a hot topic for many new families of codes, but for the ones we will be looking at we will give proofs of the value of d.

3.1 THE GENERATOR MATRIX AND PARITY CHECK MATRIX

We have just defined the generator matrix, and so given an information word of length k we can provide a codeword of length n. We need to know something further about these codewords, and the requirements placed on them. If we look at the construction of a particular codeword, c, say, then we could define another matrix H^T (T being the transpose) such that

$$\mathbf{c}H^T = 0. \tag{4.35}$$

That is a general vector of length n is a codeword only if it satisfies this equation. H is known as the parity check matrix. We can obviously say further that

$$GH^T = 0. \tag{4.36}$$

Sice all the rows of G are all linearly independent, we can by elementary row operations and column permutations reduce it to the form

$$G = [I|A]. \tag{4.37}$$

I here is the identity matrix of size $(k \times k)$.

This form is known as systematic as when the transformation $\mathbf{c} = \mathbf{u}G$ is performed the $\mathbf{c}$ contains an exact copy of $\mathbf{u}$ in it. Now this will be a generator matrix for the same code but the information word to codeword transformation has changed. It can shown that is the generator matrix, G, of a given code C, is in systematic form then the parity check matrix is of the form

$$H = [A^T|I]. \tag{4.38}$$

I here is the identity matrix of size $(n - k) \times (n - k)$.

We need now to look at an example to clarify all that has been explained.

EXAMPLE 4.38   Consider the set of binary vectors $(1, 1, 0, 1, 0, 1)$, $(0, 1, 0, 0, 1, 1)$ and $(1, 1, 1, 0, 0, 1)$ It is easy to check that this set of vectors is linearly independent. We can therefore consider that the matrix

$$G = \begin{bmatrix} 1 & 1 & 0 & 1 & 0 & 1 \\ 0 & 1 & 0 & 0 & 1 & 1 \\ 1 & 1 & 1 & 0 & 0 & 1 \end{bmatrix} \tag{4.39}$$

is a generator matrix for a $(6, 3)$ linear code. For this simple example we will look at every codeword to see what the code looks like. In Table 4.6 we see the information word and the corresponding codeword for this given generator matrix G.

So from this we can see that the minimum distance of the code is 3, so we have a $(6, 3, 3)$ linear block code. We can perform row operations on G to put it into systematic form. One form is

$$G_{(S)} = \begin{bmatrix} 1 & 0 & 0 & 1 & 1 & 1 \\ 0 & 1 & 0 & 0 & 1 & 1 \\ 0 & 0 & 1 & 1 & 0 & 1 \end{bmatrix} \tag{4.40}$$

Information Word	Codeword
000	000000
001	111000
010	010011
011	101011
100	110101
101	001100
110	100111
111	011110

Table 4.6.   Encoding procedure for Example 4.38

therefore we can obtain a parity check matrix as

$$H_{(S)} = \begin{bmatrix} 1 & 0 & 1 & 1 & 0 & 0 \\ 1 & 1 & 0 & 0 & 1 & 0 \\ 1 & 1 & 1 & 0 & 0 & 1 \end{bmatrix}, \tag{4.41}$$

and it can easily checked that $G_{(S)}H^T = 0$, and that for every codeword, c, in C, $cH^T = 0$. Of course we can construct codes by defining a generator matrix G or a parity check matrix H. So c is a codeword if and only if , $cH^T = 0$.

THEOREM 4.39 *The least number of linearly independent columns of the parity check matrix of a linear code, C, is the minimum distance of that code.*

PROOF. Proof to do. See MacWilliams and Sloane □

We will now give some more information on linear codes in general. Further to our statement on linearity before we can see now that if c_1 and c_2 are codewords then so is $c_1 + c_2$. This is easy to see from our definition of a vector being a codeword if and only if $cH^T = 0$.

$$(c_1 + c_2)H^T = c_1H^T + c_2H^T = 0 + 0 = 0. \tag{4.42}$$

Let us now look at some bounds that exist on the values of n, k and d.

THEOREM 4.40 *Singleton Bound: If C is an (n, k, d) code then $n-k \le d-1$.*

PROOF. The rank of H is $n - k$, so there must be maximum $n - k$ linearly independent columns. □

THEOREM 4.41 *Gilbert-Varshomov Bound: There exist a binary linear code of length n, with at most $n - k$ parity checks and minimum distance at least d if*

$$1 + \binom{n-1}{1} + \cdots + \binom{n-1}{d-2} < 2^{n-k} \tag{4.43}$$

PROOF. See [212]. □

Codes in which $n - k = d - 1$ are called maximum distance separable (MDS). These codes are sought after as they provide for a given n and k the largest possible d.

THEOREM 4.42 *Sphere packing/Hamming Bound: A t-error correcting binary code of length n containing 2^k codewords must satisy*

$$2^k \left(1 + \binom{n}{1} + \binom{n}{2} + \cdots + \binom{n}{t} \right) \le 2^n \qquad (4.44)$$

This can also be applied to fields of q elements to give

$$q^k \left(1 + (q-1) \binom{n}{1} + (q-1)^2 \binom{n}{2} + \cdots + (q-1)^t \binom{n}{t} \right) \le q^n$$
$$(4.45)$$

PROOF. (binary case) The spheres of radius t in n dimensional space are disjoint. Each of the 2^k spheres contain $1 + \binom{n}{1} + \cdots + \binom{n}{t}$ vectors. The total number of vectors is 2^n so we have

$$2^k \left(1 + \binom{n}{1} + \binom{n}{2} + \cdots + \binom{n}{t} \right) \le 2^n \qquad (4.46)$$

□

If the spheres of radius t around the codewords are disjoint and contain all the vectors of length n then the t-error correcting code can correct all errors of weight (t and none of weigth $> t$. A code that has this property is called a perfect code.

3.2   HAMMING CODES

Considering only binary codes, if we have a parity check matrix with r rows, then the code will have r parity checks. Also there are only $2^r - 1$ possible columns we can't have the zero column). For example if we want a code with 3 parity checks, then $r = 3$ and we have $2^3 - 1$ possible columns, i.e.

$$\begin{matrix} 1 & 0 & 1 & 0 & 1 & 0 & 1 \\ 0 & 1 & 1 & 0 & 0 & 1 & 1 \\ 0 & 0 & 0 & 1 & 1 & 1 & 1 \end{matrix} \qquad (4.47)$$

If we define this structure as H then we have a possible code of length 7 and dimension $7 - 3 = 4$.

It is easy to see that there are a maximum of three independent columns so the maximum distance of the code is 3. By varying the number of parity checks we have therefore defined a family of codes called Hamming codes. Their parameters are

$$(n, k, d) = (2^r - 1, 2^r - 1 - r, 3) \tag{4.48}$$

for any given $r \geq 2$.

So let us look again at the $(7, 4, 3)$ Hamming Code. The parity check matrix is

$$H = \begin{bmatrix} 1 & 0 & 1 & 0 & 1 & 0 & 1 \\ 0 & 1 & 1 & 0 & 0 & 1 & 1 \\ 0 & 0 & 0 & 1 & 1 & 1 & 1 \end{bmatrix} \tag{4.49}$$

If we want to find a generator matrix for this code we need to simply swap columns 1 and 5, 2 and 6 and 4 and 7 to get

$$H_{(S)} = \begin{bmatrix} 1 & 0 & 1 & 1 & 1 & 0 & 0 \\ 0 & 1 & 1 & 1 & 0 & 1 & 0 \\ 1 & 1 & 0 & 1 & 0 & 0 & 1 \end{bmatrix} \tag{4.50}$$

then we have $H_{(S)} = [A|I]$ and so can form $G_{(S)} = [I|A^T]$, i.e.

$$G_{(S)} = \begin{bmatrix} 1 & 0 & 0 & 0 & 1 & 0 & 1 \\ 0 & 1 & 0 & 0 & 0 & 1 & 1 \\ 0 & 0 & 1 & 0 & 1 & 1 & 0 \\ 0 & 0 & 0 & 1 & 1 & 1 & 1 \end{bmatrix} \tag{4.51}$$

It is trivial to show that $G_{(S)}H_{(S)T} = 0$ so we a correct $G_{(S)}$ for the $H_{(S)}$. If we wanted a generator matrix for the original parity matrix, H, then we need to apply the column permutations we used to put the parity check matrix in systematic form. We can thus get

$$G = \begin{bmatrix} 1 & 0 & 0 & 1 & 1 & 0 & 0 \\ 0 & 1 & 0 & 1 & 0 & 1 & 0 \\ 1 & 1 & 1 & 0 & 0 & 0 & 0 \\ 1 & 1 & 0 & 1 & 0 & 0 & 1 \end{bmatrix} \tag{4.52}$$

We can once again check that $GH^T = 0$.

So we have a class of single error correcting codes.

THEOREM 4.43 *Hamming Codes are perfect codes.*

PROOF. The spheres around each codeword are disjoint and of radius 1 as it is a single error correcting code. There are $n + 1$ vectors in each sphere (the original one and another n of distance one from the original). There are 2^k spheres. So in total we have

$$(n + 1)2^k = ((2^{n-k} - 1) + 1)2^k = 2^n \tag{4.53}$$

which is all the vectors possible so the code is perfect. ☐

3.3   DECODING LINEAR CODES

Let C be an (n, k, d) code over F_q. For any vector $\mathbf{v}$, the set

$$\mathbf{v} + C = \{\mathbf{v} + \mathbf{x} : \mathbf{x} \in C\} \tag{4.54}$$

is called a coset of C. The following are properties of a coset:

1. every vector $\mathbf{v}$ is in a coset,

2. if $\mathbf{v}_1$ and $\mathbf{v}_2$ are in the same coset then $(\mathbf{v}_1 - \mathbf{v}_2) \in C$,

3. each coset contains q^k vectors

4. there are q^{n-k} cosets, as each coset is unique.

Consider now that we encode a vector $\mathbf{v}$ into a codeword $\mathbf{c}$ and transmit it across a symmetric channel. The received vector is such that

$$\mathbf{r} = \mathbf{c} + \mathbf{e} \tag{4.55}$$

where $\mathbf{e}$ is the error vector, the positions and values of the errors.

Now $\mathbf{r}$ must belong to one coset, say the coset $\mathbf{v}_i + C$, so

$$\mathbf{r} = \mathbf{v}_i + \mathbf{c}_i \quad (\mathbf{c}_i \in C) \tag{4.56}$$

If the encoded word $\mathbf{c}$ was transmitted, the error vector is

$$\mathbf{e} = \mathbf{r} - \mathbf{c} = \mathbf{v}_i + \mathbf{c}_i - \mathbf{c} = \mathbf{v}_i + \mathbf{c}_j \in \mathbf{v}_i + C \tag{4.57}$$

Therefore the possible error vectors are precisely those vectors in the coset containing $\mathbf{r}$.

So if we want to decode $\mathbf{r}$, we need to choose a minimum weight vector $\hat{\mathbf{e}}$ in the coset containing $\mathbf{r}$, and then

$$\hat{\mathbf{c}} = \mathbf{r} - \hat{\mathbf{r}}. \tag{4.58}$$

This minimum weight vector is called a coset leader.

We now construct a table that will enable us to decode a received word. This table is known as the standard array. The table is constructed as follows:

1. the first row consists of the zero vector and then the coset with leader 0, i.e. the code itself

2. the other rows are the other cosets arranged in the same order as 1 with the coset leader first

When $\mathbf{r}$ is received, the position in the standard array is found. $\hat{\mathbf{e}}$ is taken to be the coset leader of the coset that contains $\mathbf{r}$, then $\hat{\mathbf{c}} = \mathbf{r} - \hat{\mathbf{e}}$ and the information word can then be retrieved.

EXAMPLE 4.44 Consider the $(5, 2, 3)$ linear code generated by the generator matrix:

$$G = \begin{bmatrix} 1 & 0 & 1 & 1 & 0 \\ 0 & 1 & 0 & 1 & 1 \end{bmatrix} \tag{4.59}$$

The standard array is

coset leader			
00000	10110	01011	11101
10000	00110	11011	01101
01000	11110	00011	10101
00100	10010	01111	11001
00010	10100	01001	11111
00001	10111	01010	11100

Table 4.7.

Let us receive $\mathbf{r} = 11110$. This is in the coset with leader 01000 so we choose

$$\begin{aligned} \mathbf{c} &= 11110 - 010000 \\ &= 10110 \end{aligned} \tag{4.60}$$

which we decode to 10.

But in general how do we find which coset the received vector is in, and do we need to construct a table at all? The answer turns out to be no.

SYNDROME DECODING

Define the vector $\mathbf{s}$ such that

$$\mathbf{s} = \mathbf{r}H^T \tag{4.61}$$

to be the syndrome of $\mathbf{r}$. It is easy to see that if the received vector $\mathbf{r}$ contains no errors then $\mathbf{s} = \mathbf{0}$, but the problem still lies in the fact that if more errors than the code can handle occur then the syndrome could also be $\mathbf{0}$.

For binary codes the syndrome is the sum of the columns where the errors occurred.

EXAMPLE 4.45 For

$$G = \begin{bmatrix} 1 & 0 & 1 & 1 & 0 \\ 0 & 1 & 0 & 1 & 1 \end{bmatrix}, H = \begin{bmatrix} 1 & 0 & 1 & 0 & 0 \\ 1 & 1 & 0 & 1 & 0 \\ 0 & 1 & 0 & 0 & 1 \end{bmatrix},$$

let us take $\mathbf{r} = 11110$, then $\mathbf{s} = \begin{bmatrix} 0 \\ 1 \\ 1 \end{bmatrix}$. We know that at least one error

occurred. If we assume that only one error occurred then the error occurred in position 2. We thus have

$$
\begin{aligned}
\mathbf{c} &= \mathbf{r} - \mathbf{e} \\
&= 11110 - 01000 \\
&= 10110
\end{aligned}
\tag{4.62}
$$

and this decodes to 10 as before.

Let us now use the syndrome decoding method to decode the $(7, 4, 3)$ Hamming Code.

EXAMPLE 4.46 We have

$$
H = \begin{bmatrix} 1 & 0 & 1 & 0 & 1 & 0 & 1 \\ 0 & 1 & 1 & 0 & 0 & 1 & 1 \\ 0 & 0 & 0 & 1 & 1 & 1 & 1 \end{bmatrix}
\tag{4.63}
$$

and

$$
G = \begin{bmatrix} 1 & 0 & 0 & 1 & 1 & 0 & 0 \\ 0 & 1 & 0 & 1 & 0 & 1 & 0 \\ 1 & 1 & 1 & 0 & 0 & 0 & 0 \\ 1 & 1 & 0 & 1 & 0 & 0 & 1 \end{bmatrix}
\tag{4.64}
$$

Let us take an information word $\mathbf{u} = 0110$. We encode this to $\mathbf{c} = 1011010$. Assume this is sent across a channel and it is received as $\mathbf{r} = 1011110$. To decode this we will use the syndrome decoding method. So

$$
\begin{aligned}
\mathbf{s} &= \mathbf{r}H^T \\
&= \begin{bmatrix} 1 \\ 0 \\ 1 \end{bmatrix}
\end{aligned}
\tag{4.65}
$$

this indicates that an error has occurred in position 5, and so $\hat{\mathbf{e}} = 0000100$. We therefore have

$$
\begin{aligned}
\hat{\mathbf{c}} &= \mathbf{r} - \hat{\mathbf{e}} \\
&= 1011110 - 0000100 \\
&= 1011010
\end{aligned}
\tag{4.66}
$$

as expected, and this then decodes to 0110.

4. CYCLIC CODES

If we consider a codeword of an (n, k, d) code to be $\mathbf{c} = (c_0, c_1, \ldots, c_{n-1})$ then the cyclic shift (right) is $\mathbf{c'} = (c_{n-1}, c_0, c_1, \ldots, c_{n-2})$. The left shift is similar. We shall consider only right shifts in this book.

DEFINITION 4.47 Consider an (n, k, d) linear code. If every cyclic shift of a codeword is also a codeword then the code is known as a *cyclic code*.

If we express the codeword as a polynomial, for example the above codeword would be $\mathbf{c}(x) = c_0 + c_1 x + \cdots + c_{n-1} x^{n-1}$, then the cyclic shift is given simply as $\mathbf{c'}(x) = c_{n-1} + c_0 x + c_1 x^2 + \cdots + c_{n-2} x^{n-1}$. It is easy to see that $\mathbf{c'}(x) = x\mathbf{c}(x) \bmod x^n - 1$.

DEFINITION 4.48 Further to Definition 4.47, a cyclic (n, k, d) code is an ideal of the polynomial ring $F_q[x] \bmod x^n - 1$.

By Theorem 4.21, every ideal of the polynomial ring $F_q[x] \bmod x^n - 1$ is a principal ideal.

The following theorem will provide the basis for linking cyclic codes and ideals of polynomial rings.

THEOREM 4.49 *Let $g(x)$ be a divisor of $x^n - 1$ over F_q. Further let $g(x)$ be monic and be of degree $n - k$. Then $g(x)$ generates (using the notation described above) a cyclic code of dimension k.*

PROOF. See [207]. $\square$

DEFINITION 4.50 For a cyclic (n, k, d) code with generator polynomial $g(x)$, the parity check polynomial $h(x)$ is given by

$$x^n - 1 = g(x)h(x) \tag{4.67}$$

We will now explain briefly the encoding procedure when considering the polynomial approach to viewing cyclic codes. In a short while we will look at manipulating cyclic codes in the generator matrix representation we are now used to.

Suppose we have a cyclic code with generator polynomial

$$g(x) = g_0 + g_1 x + g_2 x^2 + \cdots + g_{n-k-1} x^{n-k-1}. \tag{4.68}$$

Now consider the information word

$$u(x) = u_0 + u_1 x + u_2 x^2 + \cdots + u_{k-1} x^{k-1}. \tag{4.69}$$

We can simple find the polynomial $c(x) = u(x)g(x)$ as there is a 1-1 mapping from the $u(x)$ onto the $c(x)$. This method is non-systematic. For most purposes it is useful to have a systematic encoding procedure. The requirement for this stems from many facts but mainly includes the fact that if the codeword cannot be decoded to an information word then the information word part is simply passed to the user.

We can simply systematically encode the information word by taking the information word and placing it in the high order coefficients and add the parity bits as the low order coefficients. So firstly we need

$$c_s(x) = u(x)x^{n-k} \tag{4.70}$$

then we need to work out the parity bits. This is done by obtaining a $b(x)$ such that

$$u(x)x^{n-k} + b(x) = 0 \bmod g(x) \tag{4.71}$$

i.e.

$$b(x) = -u(x)x^{n-k} \bmod g(x) \tag{4.72}$$

and then

$$c(x) = c_s(x) + b(x). \tag{4.73}$$

EXAMPLE 4.51 Consider the polynomial $g(x) = x^3 + x^2 + 1$. We can generate a $(7, 4, d)$ cyclic code over F_2 using this polynomial, as $x^3 + x^2 + 1$ divides $x^7 - 1$. If we take the information polynomial as $u(x) = x^3 + x + 1$. To encode this in the (systematic) manner described above we do the following:

1. $x^{7-4}u(x) = x^6 + x^4 + x^3$

2. Now we work out $b(x)$.

$$\begin{aligned} b(x) &= -x^{7-4}u(x) \bmod g(x) \\ &= x^2 \end{aligned} \tag{4.74}$$

3. $c(x) = x^6 + x^4 + x^3 + x^2$

and so we have the codeword as $x^6 + x^4 + x^3 + x^2$.

Of course we can encode this in the non-systematic way by simply multiplying the information polynomial to the generator polynomial. We thus get

$$\begin{aligned} c(x) &= u(x)g(x) \\ &= x^6 + x^5 + x^3 + x^4 + x^3 + x^2 + x + 1. \end{aligned} \tag{4.75}$$

4.1   MATRIX NOTATION OF CYCLIC CODES

For exemplary reasons we will now look at cyclic codes using the familiar matrix notation. Let the generator polynomial of the cyclic code be

$$g(x) = g_0 + g_1 x + g_2 x^2 + \cdots + g_{n-k-1} x^{n-k-1}. \tag{4.76}$$

The code polynomials are of the form

$$
\begin{aligned}
c(x) &= u(x)g(x) \\
&= u_0 g(x) + u_1 g(x)x + u_2 g(x)x^2 + \cdots + u_{k-1} g(x) x^{k-1}.
\end{aligned} \tag{4.77}
$$

and there are q^k different code polynomials. We can put the coefficients of the code polynomial in vector form as follows.

$$
G = \begin{bmatrix}
g_0 & g_1 & g_2 & \cdots & g_{n-k} & 0 & 0 & \cdots & 0 & 0 \\
0 & g_0 & g_1 & \cdots & g_{n-k-1} & g_{n-k} & 0 & \cdots & 0 & 0 \\
\vdots & & \ddots & & & & \ddots & & & \\
0 & 0 & 0 & \cdots & g_0 & g_1 & g_2 & \cdots & g_{n-k} & 0 \\
0 & 0 & 0 & \cdots & 0 & g_0 & g_1 & \cdots & g_{n-k-1} & g_{n-k}
\end{bmatrix}
$$

$$
= \begin{bmatrix}
G_0 \\
G_1 \\
\vdots \\
G_{k-1}
\end{bmatrix}
$$

$$\tag{4.78}$$

This matrix obviously has all the rows independent, so the linear combinations of the $G_i, i = 0, \ldots, k-1$ form a k-dimensional subspace of n- dimensional space over F_q. Therefore this is indeed the generator matrix of an (n, k, d) code. We of course do not know d yet.

We now wish to look at the parity check matrix, H, of a cyclic code, C. We are working within the ring $F_q/(f(x)$, where $f(x) = x^n - 1)$, and have the generator polynomial of the cyclic code as $g(x)$.

Consider $h(x) = f(x)/g(x)$. Now degree $g(x) = n - k - 1$ so degree $h(x) = k - 1$. Further $g(x)$ is monic so $h(x)$ is monic. It follows that $h(x)$ generates a cyclic code C' of dimension $n - k$.

We need to show that C' is indeed the dual code of the cyclic code C. Consider a codeword, $c(x) = u(x)f(x)$, in C, then multiplying this by $h(x)$ we get

$$
\begin{aligned}
c(x)h(x) &= u(x)g(x)h(x) \\
&= u(x)f(x) \\
&= u(x) + x^n u(x).
\end{aligned} \tag{4.79}
$$

Since the degree of $u(x) \le k - 1$, the coefficients of $x^k, x^{k+1}, \ldots, x^{n-1}$ are zero in the final line of 4.79. We expand the equation on the left hand side and equate the coefficients for degrees $k, \ldots, n-1$ to zero. We thus get

$$\sum_{i=0}^{k} c_{k-i} h_i = 0$$

$$\sum_{i=0}^{k} c_{k+1-i} h_i = 0 \tag{4.80}$$

$$\vdots$$

$$\sum_{i=0}^{k} c_{n-i} h_i = 0.$$

Consider the $((n-k) \times n)$ matrix

$$H = \begin{bmatrix} h_{n-k} & h_{n-k-1} & h_{n-k-2} & \cdots & h_0 & 0 & 0 & \cdots & 0 \\ 0 & h_{n-k} & h_{n-k-1} & \cdots & h_1 & h_0 & 0 & \cdots & 0 \\ \vdots & & \ddots & & & & & \ddots & \\ 0 & 0 & \cdots & h_{n-k} & h_{n-k-1} & \cdots & h_1 & h_0 & 0 \\ 0 & 0 & \cdots & 0 & h_{n-k} & h_{n-k-1} & \cdots & h_1 & h_0 \end{bmatrix} \tag{4.81}$$

It is easily seen that if we consider the polynomial $x^k h(x^{-1})$ then we get the above matrix as the generator matrix of a $(n, n-k, d)$ code. It follows from 4.80 that any codeword $c \in C$ is orthogonal to every row of H. Therefore H is a parity check matrix of the cyclic code C.

DEFINITION 4.52 The polynomial $h(x)$ used to form the parity check matrix, H, of a cyclic code C, is known as the *parity check polynomial*.

We have thus proved the following theorem.

THEOREM 4.53 *Let C be an (n, k, d) cyclic code generated by the polynomial $g(x)$. Then the dual code of C is also cyclic and is generated by the polynomial $h(x) = f(x)/g(x)$, where $f(x) = x^n - 1$.*

EXAMPLE 4.54 Consider again the polynomial $g(x) = x^3 + x^2 + 1$. We can once again generate a $(7, 4, d)$ cyclic code over F_2 using this polynomial, as $x^3 + x^2 + 1$ divides $x^7 - 1$. We will this time however develop the generator matrix. We have $g(x)$ so the generator matrix is

$$G = \begin{bmatrix} 1 & 0 & 1 & 1 & 0 & 0 & 0 \\ 0 & 1 & 0 & 1 & 1 & 0 & 0 \\ 0 & 0 & 1 & 0 & 1 & 1 & 0 \\ 0 & 0 & 0 & 1 & 0 & 1 & 1 \end{bmatrix}. \tag{4.82}$$

If we take the information polynomial as $u(x) = x^3 + x + 1$ then this corresponds to the information vector $\mathbf{u} = [1101]$. We encode this in the usual (non-systematic) manner as

$$\mathbf{c} = \mathbf{u}G \tag{4.83}$$

and we get $\mathbf{c} = [1111111]$, as expected and given in Example 4.51. To encode this in a systematic form we need to find a new systematic generator matrix for the cyclic code. Therefore we need to perform elementary row operations on (4.82).

We can thus obtain the (systematic) generator matrix as

$$G = \begin{bmatrix} 1 & 0 & 1 & 1 & 0 & 0 & 0 \\ 1 & 1 & 1 & 0 & 1 & 0 & 0 \\ 1 & 1 & 0 & 0 & 0 & 1 & 0 \\ 0 & 1 & 1 & 0 & 0 & 0 & 1 \end{bmatrix}. \tag{4.84}$$

Now we encode the information word $\mathbf{u} = [1101]$ and we get

$$\begin{aligned} \mathbf{c} &= \mathbf{u}G \\ &= [0011101], \end{aligned} \tag{4.85}$$

which corresponds to the information polynomial $c(x) = x^6 + x^4 + x^3 + x^2$ as expected and seen in Example 4.51.

We now want to consider the parity check matrix for this code.

Having G in systematic form it is easy to form an H and we get a parity check matrix as

$$H = \begin{bmatrix} 1 & 0 & 0 & 1 & 1 & 1 & 0 \\ 0 & 1 & 0 & 0 & 1 & 1 & 1 \\ 0 & 0 & 1 & 1 & 1 & 0 & 1 \end{bmatrix}. \tag{4.86}$$

It is easy to see (as expected) that

$$H\mathbf{c}^T = \mathbf{0}, \tag{4.87}$$

and further that

$$HG^T = \mathbf{0}. \tag{4.88}$$

4.2 DECODING OF CYCLIC CODES

Following on from the decoding of linear codes in Section 3.3, we can see that all the ideas there hold true for cyclic codes.

SYNDROME DECODING

We know that

$$\mathbf{s} = H\mathbf{r}^T \tag{4.89}$$

where

$$\mathbf{r} = \mathbf{c} + \mathbf{e} \tag{4.90}$$

and **s** is the syndrome vector. As before we know that

$$H\mathbf{v}^T = \mathbf{0} \tag{4.91}$$

so

$$\mathbf{s} = H\mathbf{e}^T. \tag{4.92}$$

When considering cyclic codes we want to work with polynomials. If the code is an (n, k) systematic code generated by $g(x)$, then we can represent the syndrome **s** as a polynomial $s(x)$ of degree at most $(n - k - 1)$. We will now show that $s(x)$ can be found by summing the received parity polynomial and the parity recomputed from the information part of the received polynomial. This is essentially the polynomial approach to 4.89. Let $r(x)$ be the received word in polynomial form. This will have degree at most $(n - 1)$. We know that

$$r(x) = c(x) + e(x) \tag{4.93}$$

where $c(x) = c_s(x) + b(x)$ is the original codeword to be trannsmitted and $e(x)$ can be decomposed into

$$e(x) = e_s(x) + e_p(x). \tag{4.94}$$

Then

$$\begin{aligned} r(x) &= c_s(x) + e_s(x) + b(x) + e_p(x) \\ &= c'_s(x) + b'(x). \end{aligned} \tag{4.95}$$

Now divide $r(x)$ by $g(x)$ to get

$$r(x) = q(x)g(x) + s(x) \tag{4.96}$$

where $q(x)$ is the quotient and $s(x)$ is the remainder. Substituting 4.95 into 4.95 we get

$$c'_s(x) + b'(x) = q(x)g(x) + s(x) \tag{4.97}$$

i.e.

$$\frac{c'_s(x)}{g(x)} + \frac{b'(x)}{g(x)} = q(x) + \frac{s(x)}{g(x)}. \tag{4.98}$$

Now taking remainders of 4.98 or equivalently taking 4.96 modulo $g(x)$, we get

$$s(x) = b'(x) + c_s(x) \bmod g(x). \tag{4.99}$$

Now we know that

$$\begin{aligned} q(x)g(x) + s(x) &= c(x) + e(x) \\ &= m(x)c(x) + e(x) \end{aligned} \tag{4.100}$$

as $c(x)$ is a multiple of $g(x)$, and

$$e(x) = (q(x) - m(x))g(x) + s(x) \qquad (4.101)$$

or equivalently

$$s(x) \equiv e(x) \bmod g(x). \qquad (4.102)$$

EXAMPLE 4.55 Consider a $(7, 4)$ cyclic code generated by $g(x) = x^3 + x^2 + 1$. We need to consider the syndrome polynomials and which errors they are assciated with. We form the following table from the parity check matrix which we met in Example 4.54. Now we have the problem set up we can decode a

$s(x)$	Error position
1	0
x	1
$x + 1$	5
x^2	2
$x^2 + 1$	3
$x^2 + x$	6
$x^2 + x + 1$	4

Table 4.8. Syndromes with corresponding error position for Example 4.55

code polynomial.

Consider the information polynomial $u(x) = x^6 + x^4 + x^3$, this encodes (as we have seen in Example 4.51) to $c(x) = x^6 + x^4 + x^3 + x^2$. Now consider that an error occurred and in fact we received $r(x) = x^6 + x^5 + x^4 + x^3 + x^2$. We know that there is an error in position 5, and in fact the coefficient fo x^5 should be zero. We will go through the full decoding procedure to correctly decode this. We know that $s(x) \equiv r(x) \bmod g(x)$, so

$$
\begin{aligned}
s(x) &= x^6 + x^5 + x^4 + x^3 + x^2 \bmod x^3 + x^2 + 1 \\
&= x + 1.
\end{aligned} \qquad (4.103)
$$

This corresponds to position 5 in the received polynomial and we thus get the estimate of the correct coded polynomial as

$$\hat{c}(x) = x^6 + x^4 + x^3 + x^2, \qquad (4.104)$$

and therefore the information polynomial (the systematic part of the correctly coded polynomial) is

$$u(x) = x^6 + x^4 + x^3, \qquad (4.105)$$

as expected.

5.   REED-SOLOMON CODES

It is customary to initially consider binary narrow sense primitive BCH codes [198, 199, 203] which are a superset of the Reed-Solomon codes [202]. However as the main block codes used in broadcasting are Reed-Solomon codes, we will concentrate on these and refer the reader to one of the many standard text books on coding theory for an explanation of BCH codes [203, 205].

If for the minute we look at non-binary BCH codes, then for any choice of positive integers s and t there exists a BCH code over F_q of length $n = q^s - 1$, which corrects t or fewer errors with $n - k \ leq 2st$.

We learnt about cyclic codes in the previous sections and we take this a step further now. Let α be a primitive element in F_{q^s}. The generator polynomial $g(x)$ of a t-error correcting BCH code over F_q is the polynomial of least degree with coefficients from F_q with roots $\alpha, \alpha^2, \ldots \alpha^{2t}$. Therefore

$$g(x) = \mathrm{lcm}\{m_1(x), m_2(x), \ldots, m_{2t}(x)\} \tag{4.106}$$

where $m_i(x)$ is the minimal polynomial of α^i. The degree of $g(x)$ is less than or equal to $2st$ (as each $m_i(x)$ has degree less than or equal to s), hence by Theorem 4.49 there are less than or equal to $2st$ parity checks.

If $q = 2$ we get binary BCH codes. If $s = 1$ we get a subclass of BCH codes over F_q that are known as Reed-Solomon Codes or RS Codes [202].

RS Codes have the following parameters

Block Length	$n = q - 1$
Number of parity checks	$n - k = 2t$
Minimum Distance	$d_{min} = 2t + 1$

We need to prove that the actual minimum distance of the RS code is equal to the designed distance. From the Singleton Bound 4.40 we have that

$$d_{min} \leq n - k + 1 \tag{4.107}$$

and from the BCH Bound [212] we have

$$d_{min} \geq n - k + 1 \tag{4.108}$$

therefore

$$\begin{aligned} d_{min} &= n - k + 1 \\ &= 2t + 1. \end{aligned} \tag{4.109}$$

Since an RS code is simply a cyclic code, if we can find a generator polynomial we can define the entire encoding procedure.

Consider the symbols of the code as being over F_{q^m}. Let α be a primitive element in F_{q^m}. We will define a (primitive) RS code. If we choose a non-primitive element then we could form a (non-primitive) RS code. Let us require that the RS code has an error correcting capability of t errors. Then the generator polynomial is defined as

$$g(x) = (x - \alpha)(x - \alpha^2) \cdots (x - \alpha^{2t}). \tag{4.110}$$

All the prooperties that we developed in the previous section can be applied to RS codes.

5.1 ENCODING REED-SOLOMON CODES

Consider an information word to be

$$u(x) = u_0 + u_1 x + u_2 x^2 + \cdots + u_{k-1} x^{k-1} \tag{4.111}$$

with the u_i, $i = 0, \ldots, k - 1$ are from F_{q^m}. Then as we have already seen this can be (systematically) encoded by the process

1. Work out the systematic part, $c_s(x) = x^{n-k} u(x)$.

2. Now work out $b(x)$ where

$$b(x) = -x^{n-k} u(x) \bmod g(x) \tag{4.112}$$

   i.e. $b(x)$ is the result from dividing the message polynomail multiplied by x^{n-k} by the generator polynomial $g(x)$.

3. Then $c(x) = c_s(x) + b(x)$.

We wil give a substantial example in the next section that will show encoding and decoding of RS decoding.

5.2 REED-SOLOMON DECODERS

There are many ways to decode RS codes, some more efficient than others. We shall see in the section one very efficient and generally understood as one of the best algorithms in terms of implementability. We will start by looking at the problem in general and then give a full explanation of the efficient algorithm that we have chosen for this book. It must be noted that for certain implementation issues the methods we explain below are perhaps not the best.

A thorough explanation of many other algorithms and methods can be found in [212, 205, 203, 213].

We will now explain the situation in the most general terms possible. Let the received word be

$$r(x) = r_0 + r_1 x + r_2 x^2 + \cdots + r_{n-1} x^{n-1} \tag{4.113}$$

and let the transmitted word be (which obviously we do not know at the decoder)

$$c(x) = c_0 + c_1 x + c_2 x^2 + \cdots + c_{n-1} x^{n-1}. \tag{4.114}$$

Now $c(x)$ is a valid codeword polynomial, and in general $r(x)$ is not. The error is then given by

$$
\begin{aligned}
e(x) &= r(x) - c(x) \\
&= e_0 + e_1 x + e_2 x^2 + \cdots + e_{n-1} x^{n-1}.
\end{aligned}
\tag{4.115}
$$

Assume that $\nu \le t$ errors have occurred (i.e. we are using the code up to its full error correcting capability), and so there will be ν non-zero terms in 4.115. Let the positions that these non-zero terms appear at be $i_1, i_2, \ldots, i_\nu$, with $i_l < i_{l+1}$.

We calculate the syndromes, S_i as normal, i.e.

$$
\begin{aligned}
S_i &= e(\alpha^i) \\
&= r(\alpha^i)
\end{aligned}
\tag{4.116}
$$

From 4.115 and 4.116 we get the syndromes as

$$
\begin{aligned}
S_1 &= e_{i_1} \alpha^{i_1} + e_{i_2} \alpha^{i_2} + \cdots + e_{i_\nu} \alpha^{i_\nu} \\
S_2 &= e_{i_1} (\alpha^{i_1})^2 + e_{i_2} (\alpha^{i_2})^2 + \cdots + e_{i_\nu} \alpha^{i_\nu})^2 \\
&\vdots \\
S_{2t} &= e_{i_1} (\alpha^{i_1})^{2t} + e_{i_2} (\alpha^{i_2})^{2t} + \cdots + e_{i_\nu} (\alpha^{i_\nu})^{2t}
\end{aligned}
\tag{4.117}
$$

Now define

$$
\beta_l = \alpha^{i_l}, l = 0, 1, \nu - 1.
\tag{4.118}
$$

The $\beta_l, l = 1, \ldots, \nu$ are known as the *error location numbers*. We thus obtain the syndromes as

$$
\begin{aligned}
S_1 &= e_{i_1} \beta_1 + e_{i_2} \beta_2 + \cdots + e_{i_\nu} \beta_\nu \\
S_2 &= e_{i_1} \beta_1^2 + e_{i_2} \beta_2^2 + \cdots + e_{i_\nu} \beta_\nu^2 \\
&\vdots \\
S_{2t} &= e_{i_1} \beta_1^{2t} + e_{i_2} \beta_2^{2t} + \cdots + e_{i_{\nu-1}} \beta_{\nu-1}^{2t}
\end{aligned}
\tag{4.119}
$$

So any solution to 4.119 is a decoding algorithm for RS codes.

We can re-write 4.119 in matrix form as

$$
\begin{bmatrix} S_1 \\ S_2 \\ \vdots \\ S_\nu \end{bmatrix} = \begin{bmatrix} \beta_1 & \beta_2 & \cdots & \beta_\nu \\ \beta_1^2 & \beta_2^2 & \cdots & \beta_\nu^2 \\ \vdots & \vdots & \ddots & \vdots \\ \beta_1^{2t} & \beta_2^{2t} & \cdots & \beta_\nu^{2t} \end{bmatrix} \begin{bmatrix} e_{i_1} \\ e_{i_2} \\ \vdots \\ e_{i_\nu} \end{bmatrix}.
\tag{4.120}
$$

To decode the RS code we need to find a solution to 4.120. That is, we know $S_i, i = 1, \ldots, \nu$, and need to find $\beta_i, i = 1, \ldots, \nu$ and $e_{i_j}, j = 1, \ldots, \nu$. In general 4.120 has many solutions, and each soltuion will give a different error pattern. If the number of errors in the received word, $r(x)$, and therefore $e(x)$, is less than t, then the solution to 4.120 that has the least number of errors is the correct solution.

Define the following polynomial

$$\sigma(x) = \prod_{i=1}^{\nu} (1 - \beta_i x)$$
$$= \sigma_0 + \sigma_1 x + \sigma_2 x^2 + \cdots + \sigma_\nu x^\nu$$
(4.121)

so the roots of $\sigma(x)$ are the reciprocals of the error location numbers. We can see that

$$
\begin{aligned}
\sigma_1 &= 1 \\
\sigma_2 &= -(\beta_1 + \beta_2 + \cdots + \beta_\nu) \\
\sigma_3 &= (\beta_1\beta_2 + \beta_1\beta_3 + \cdots + \beta_{\nu-1}\beta_\nu \\
&\vdots \\
\sigma_\nu &= (-1)^\nu \beta_1\beta_2 \ldots \beta_\nu
\end{aligned}
$$
(4.122)

We can see that 4.119 and 4.122 are related, and indeed they are related by the following *Newton's Identities*:

$$
\begin{aligned}
S_1 + \sigma_1 &= 0 \\
S_2 + \sigma_1 S_1 + 2\sigma_2 &= 0 \\
S_3 + \sigma_1 S_2 + \sigma_2 S_1 + 3\sigma_3 &= 0 \\
&\vdots \\
S_\nu + \sigma_1 S_{\nu-1} + \cdots + \sigma_{\nu-1} S_1 + \nu\sigma_\nu &= 0 \\
S_{\nu+1} + \sigma_1 S_\nu + \cdots + \sigma_{\nu-1} S_2 + \sigma_\nu S_1 &= 0 \\
&\vdots
\end{aligned}
$$
(4.123)

We need a method to find a solution to the above Newton's Identities. This can be achieved in an iterative fashion, and the way we will describe now is known as Berlekamp's Iterative Algorithm [214, 215]. This approach is very similar to that of [216] who used an iterative procedure based on the the formation of the error location polynomial as a linear feedback shift regsiter. As such the algorithm is sometimes known as the Berlekamp-Massey Algorithm.

Let us consider the first step of the iteration. We want to find a minimum degree polynomial $\sigma^{(1)}(x)$ whose coefficients satisfy the first first line of in 4.123, i.e.

$$S_1 + \sigma_1 = 0.$$
(4.124)

Next we check whether the coefficients of $\sigma^{(1)}(x)$ also satisfy the second line of 4.123, i.e.

$$S_2 + \sigma_1 S_1 + 2\sigma_2 = 0.$$
(4.125)

If the coefficients do satisfy this then set $\sigma^{(2)}(x) = \sigma^{(1)}(x)$, otherwsie we need to add a correction term to $\sigma^{(1)}(x)$ to form $\sigma^{(2)}(x)$ such that $\sigma^{(2)}(x)$ has minimum degree and its coefficients satisfy the first two lines of 4.123, i.e. 4.124 and 4.125.

This continues on until we have reached iteration $2t$, and then we have $\sigma^{(2t)}(x)$ that satisfies $2t$ lines of 4.123. Then set $\sigma(x) = \sigma^{(2t)}(x)$. This $\sigma(x)$ will then yield a minimum weight error pattern $e(x)$ that satisfies 4.122 and therefore 4.119. However how do we add this correction term in order to enable to iterative algorithm to achieve a result?

Consider that we have reached iteration step μ, and we have a minimal degree polynomial

$$\sigma^{(\mu)}(x) = 1 + \sigma_1^{(\mu)}x + \sigma_2^{(\mu)}x^2 + \cdots + \sigma_{l_\mu}^{(\mu)}x^{l_\mu}. \qquad (4.126)$$

We want to determine $\sigma^{(\mu+1)}(x)$. To do this we need to compute a quantity, d_μ, known as the μth *discrepancy* which is defined as

$$d_\mu = S_{\mu+1} + \sigma_1^{(\mu)}S_\mu + \sigma_2^{(\mu)}S_{\mu-1} + \cdots + \sigma_{l_\mu}^{(\mu)}S_{\mu+1-l_\mu}. \qquad (4.127)$$

If $d_\mu = 0$, then the coefficients of $\sigma^{(\mu)}(x)$ satisfy the $(\mu + 1)$th Newton's identity, and we set

$$\sigma^{(\mu+1)}(x) = \sigma^{(\mu)}(x). \qquad (4.128)$$

If $d_\mu \neq 0$ then the coefficients of $\sigma^{(\mu)}(x)$ do not satisfy the $(\mu+1)$th Newton's identity, and we need to add a correction term to obtain $\sigma^{(\mu+1)}(x)$ from $\sigma^{(\mu)}(x)$. To obtain this correction term we need to consider the steps less than μ and find a $\sigma^{(\rho)}(x)$ such that the ρth discrepancy, d_ρ is non-zero and $\rho - l_\rho$ has the largest value. Then we calculate $\sigma^{(\mu+1)}(x)$ as

$$\sigma^{(\mu+1)}(x) = \sigma^{(\mu)} + d_\mu d_\rho^{-1}x^{\mu-\rho}\sigma^{(\rho)}(x). \qquad (4.129)$$

Now 4.129 is the minimum degree polynomial whose coefficients satisfy the first $\mu+1$ Newton's identities of 4.123. For the proof of this see [214, 215, 216].

So at this stage we have an algorithm that will give the error locator polynomial. The roots of $\sigma(x)$ can be found by simply substituting $1, \alpha, \alpha^2, \ldots, \alpha^{n-1}$ into $\sigma(x)$. As the roots of $\sigma(x)$ are the reciprocals of the error position if α^i is a root then α^{n-i} is an error location number, and the received polynomial 4.113 is in error at position $n - i$, so r_{n-i} is incorrect. However at this stage (and as we are considering possibly non-binary codes) we do not know the value of the errors. We therefore now need to consider finding the error values.

Let us define the a new polynomial that wil enable us to find the error values. Define

$$\begin{aligned} Z(x) = \ &1 + (S_1 + \sigma_1)x + (S_2 + \sigma_1 S_1 + \sigma_2)x^2 \\ &+ \cdots \\ &+ (S_\nu + \sigma_1 S_{\nu-1} + \sigma_2 S_{\nu-2} + \cdots + \sigma_\nu)x^\nu. \end{aligned} \qquad (4.130)$$

Now it can be shown [215] that the error value at location l is given by

$$e_{j_l} = \frac{Z(\beta_l^{-1})}{\prod\limits_{\substack{i=1 \\ i \neq l}}^{\nu} (1 + \beta_l \beta_i^{-1})}.$$

(4.131)

We will now give a complete example that will show the decoding in practice.

EXAMPLE 4.56 Consider a $(15, 9)$ RS code that is generated by the polynomial

$$g(x) = \alpha^6 + \alpha^9 x + \alpha^6 x^2 + \alpha^4 x^3 + \alpha^{14} x^4 + \alpha^{10} x^5 + x^6.$$

(4.132)

This is a code over F_{16}, α is a primitive element, where the field generating polynomial is

$$p(x) = x^4 + x + 1.$$

(4.133)

For completeness we give a table of the finite field F_{16} generated by $p(x)$ in Table 4.9

Element	Poly Rep
0	0
1	1
α	α
α^2	α^2
α^3	α^3
α^4	$1 + \alpha$
α^5	$\alpha + \alpha^2$
α^6	$\alpha^2 + \alpha^3$
α^7	$1 + \alpha + \alpha^3$
α^8	$1 + \alpha^2$
α^9	$\alpha + \alpha^3$
α^{10}	$1 + \alpha + \alpha^2$
α^{11}	$\alpha + \alpha^2 + \alpha^3$
α^{12}	$1 + \alpha + \alpha^2 + \alpha^3$
α^{13}	$1 + \alpha^2 + \alpha^3$
α^{14}	$1 + \alpha^3$

Table 4.9. Table of F_{16} in Polynomial Form

This code is a 3 error correcting code as $d = n - k + 1 = 15 - 9 + 1 = 7$. Consider the information word

$$u(x) = 1 + x + \alpha^2 x^3 + (1 + \alpha + \alpha^3)x^5 + (1 + \alpha^2)x^7.$$

(4.134)

We (systematically) encode this (using the normal encoding procedure for cyclic codes) as follows. Define

$$
\begin{aligned}
c^{(s)}(x) &= x^{15-9}u(x) \\
&= x^6 + x^7 + \alpha^2 x^9 + (1 + \alpha + \alpha^3)x^{11} + (1 + \alpha^2)x^{13}.
\end{aligned}
$$
(4.135)

Then we form $b(x)$ from

$$
\begin{aligned}
b(x) &= x^{15-9}u(x) \bmod p(x) \\
&= (\alpha^2 + \alpha^3) + \alpha^2 x + \\
&\quad (\alpha^2 + \alpha^3)x^2 + (\alpha + \alpha^3)x^4 + \\
&\quad (1 + \alpha + \alpha^2)x^5
\end{aligned}
$$
(4.136)

and finally

$$
\begin{aligned}
c(x) &= c^{(s)}(x) + b(x) \\
&= (\alpha^2 + \alpha^3) + \alpha^2 x + \\
&\quad (\alpha^2 + \alpha^3)x^2 + (\alpha + \alpha^3)x^4 + \\
&\quad (1 + \alpha + \alpha^2)x^5 + x^6 + x^7 + \alpha^2 x^9 + \\
&\quad (1 + \alpha + \alpha^3)x^{11} + (1 + \alpha^2)x^{13}).
\end{aligned}
$$
(4.137)

Now let us imagine we sent this across a channel and the following was received

$$
\begin{aligned}
r(x) &= (\alpha^2 + \alpha^3) + \alpha^2 x + \alpha^2 x^2 + (\alpha + \alpha^3)x^4 + \\
&\quad (1 + \alpha + \alpha^2)x^5 + (1 + \alpha^2 + \alpha^3)x^6 + x^7 + \\
&\quad \alpha^2 x^9 + (1 + \alpha + \alpha^3)x^{11} + (1 + \alpha^2)x^{13} + \\
&\quad (1 + \alpha + \alpha^2)x^{14},
\end{aligned}
$$
(4.138)

then we can easily see that the error polynomial is

$$
e(x) = \alpha^3 x^2 + (\alpha^2 + \alpha^3)x^6 + (1 + \alpha + \alpha^2)x^{14}.
$$
(4.139)

However if we do not know $e(x)$ (which is the usual case) then can we decode $r(x)$. We assume that at most three errors have occurred and we are using the code only to its maximum error correcting performance.

First of all we ned to obtain the syndromes $S_i = r(\alpha^i)$ and we get

$$
\begin{aligned}
S_1 &= 1 + \alpha \\
S_2 &= \alpha + \alpha^2 \\
S_3 &= 1 + \alpha + \alpha^3 \\
S_4 &= 1 + \alpha \\
S_5 &= 1 + \alpha + \alpha^2 \\
S_6 &= 1 + \alpha^2 + \alpha^3
\end{aligned}
$$
(4.140)

We will explain the first few steps so that the reader can fully understand the algorithm. The table starts as can be seen in Table 4.10.

μ	$\sigma^{(\mu)}(x)$	d_μ	l_μ	$\mu - l_\mu$	ρ
-1	1	1	0	-1	
0	1	S_1	0	0	

Table 4.10. Initial Table for Berlekamp's Algorithm

Of course we have not got any elements for ρ yet as we have not started the actual algorithm so these are left blank. Also we know that (in this example) $S_1 = 1 + \alpha$.

On step $\mu = 0$ to form $\sigma^{(1)}$ and d_1

We choose $\rho = -1$

$$
\begin{aligned}
\sigma^{(1)} &= \sigma^{(0)}(x) + d_\mu d_\rho^{-1} x^{\mu-\rho} \sigma^{(\rho)}(x) \\
&= 1 + (\alpha + 1)1^{-1} x^{(0-(-1))} \sigma^{(-1)}(x) \\
&= 1 + (\alpha + 1)x
\end{aligned}
\tag{4.141}
$$

and

$$
\begin{aligned}
d_1 &= S_2 + \sigma_1^{(1)} S_1 \\
&= (\alpha + \alpha^2) + (1 + \alpha)(1 + \alpha) \\
&= 1 + \alpha
\end{aligned}
\tag{4.142}
$$

On step $\mu = 1$ to form $\sigma^{(2)}$ and d_2

We choose $\rho = 0$

$$
\begin{aligned}
\sigma^{(2)} &= \sigma^{(1)}(x) + d_\mu d_\rho^{-1} x^{\mu-\rho} \sigma^{(\rho)}(x) \\
&= (1 + (\alpha + 1)x) + (1 + \alpha)(1 + \alpha)^{-1} x^{(1-0)} \sigma^{(0)}(x) \\
&= 1 + \alpha x
\end{aligned}
\tag{4.143}
$$

and

$$
\begin{aligned}
d_2 &= S_3 + \alpha \sigma_1^{(2)} S_2 \\
&= (1 + \alpha + \alpha^3) + (\alpha + \alpha^2) \\
&= 1 + \alpha + \alpha^2
\end{aligned}
\tag{4.144}
$$

Continuing with this until we have comlpeted 6 iterations we obtain Table 4.11. We therfore get the error locating polynomial as

$$
1 + x + x^2 + (1 + \alpha + \alpha^3)x^3,
\tag{4.145}
$$

and if we use the Chien [217] search technique (trial and error), by substituting $1, \alpha, \alpha^2, \ldots, \alpha^{14}$ we get the roots of this polynomial as α, α^9 and α^{13}. However these are the reciprocal of the error locations and so we get the error locators as α^2, α^6 and α^{14}, and the actual positions as 2, 6 and 14 expected.

μ	$\sigma^{(\mu)}(x)$	d_μ	l_μ	$\mu - l_\mu$	ρ
-1	1	1	0	-1	
0	1	$\alpha + 1$	0	0	
1	$1 + (\alpha + 1)x$	$1 + \alpha$	1	0	-1
2	$1 + \alpha x$	$1 + \alpha + \alpha^2$	1	1	0
3	$1 + (\alpha + \alpha^2 + \alpha^3)x + (1 + \alpha + \alpha^2)x^2$	$\alpha + \alpha^3$	2	1	1
4	$1 + (1 + \alpha + \alpha^2)x + (\alpha + \alpha^2)x^2$	1	2	2	2
5	$1 + (1 + \alpha + \alpha^3)x + \alpha x^3 + \alpha x^3$	$\alpha + \alpha^3$	3	2	3
6	$1 + x + x^2 + (1 + \alpha + \alpha^3)x^3$	–	–	–	–

Table 4.11.   Berlekamp Algorithm

We now need to obtain the error values and for this we use 4.130 and 4.131. So we get

$$\begin{aligned}
Z(x) &= 1 + (S_1 + 1)x + (S_2 + 1.S_1 + 1)x^2 + \\
&\quad (S_3 + 1.S_2 + 1.S_1 + (1 + \alpha + \alpha^3)x^3 \qquad (4.146) \\
&= 1 + \alpha x + \alpha^2 x^2 + (1 + \alpha^2)x^3.
\end{aligned}$$

Now we can obtain the error values from 4.131, i.e.

$$e_{j_l} = \frac{Z(\beta_l^{-1})}{\displaystyle\prod_{\substack{i=1 \\ i \neq l}}^{\nu} (1 + \beta_l \beta_i^{-1})}. \qquad (4.147)$$

For the position 14 error we have $\beta_{14} = \alpha^1$, and so we get

$$\begin{aligned}
e_{j_{14}} &= \frac{Z(\alpha)}{(1+\alpha^9\alpha)(1+\alpha^{13}\alpha)} \\
&= \frac{\alpha^3}{1+\alpha^2} \\
&= \frac{\alpha^3}{\alpha^8} \qquad (4.148) \\
&= \alpha^{10} \\
&= 1 + \alpha + \alpha^2.
\end{aligned}$$

Similarly we get the error values of $(\alpha^2 + \alpha^3)$ for position 6 and α^3 for position 2. We therefore get the error evaluating polynomial as

$$e(x) = \alpha^3 x^2 + (\alpha^2 + \alpha^3)x^6 + (1 + \alpha + \alpha^2)x^{14}, \qquad (4.149)$$

which is the same as 4.139.

6.   CONVOLUTIONAL CODES

A convolutional encoder can be thought of a linear device which maps a k-dimensional space input sequence to an n-dimensional output sequence. Further to this the difference between convolutional encoders and convolutional encoders is that convolutional encoders have memory, such that the output at a particular time is dependent not only on the current input but also some finite memory. Basically put they are finite state machines, where the output is determined by the input and the state of the machine.

If for each clock tick we input to the encoder a k-dimensional word, and the encoder has a memory of m states then we get Figure 4.2.

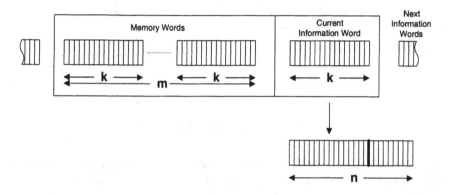

Figure 4.2.   Convolutional Encoding

We can see in Figure 4.2 that the encoding procedure is continuous in that we have a window which moves to the right. Upon each clock tick we take in a new information word and remove the last memory word. The memory held in the state machine may not be an identical version of the input, but in fact could be a function of the input and the current memory. We shall see this case when we look at systematic codes.

DEFINITION 4.57 Define an (n, k, m) convolutional code over the finite field F_q is a k-input, n-output, time-invariant, causal finite state machine of encoder memory order m.

In a simple case, we can think of a convolutional encoder as a shift register system with k inputs and n multi-input modulo q adders.

DEFINITION 4.58 The *constraint length*, ν, is defined as the number of shifts over which a single information word can influence the system.

We will concern ourselves mainly with $k = 1$ convolutional codes.

6.1   SHIFT REGISTER REPRESENTATION

We can see in Figure 4.3 an example of a $(2, 1, 3)$ convolutional encoder over F_2. This is simply a shift register.

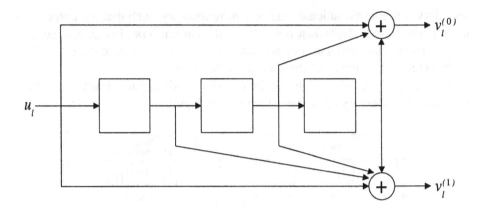

Figure 4.3.   A $(2, 1, 3)$ Convolutional Encoder

Looking at this system, for each clock tick we have one bit input, three delay units, and two bit outputs. The two bit outputs are formed from two modulo two adders, one is taken from the input at time t, $t - 2$ and $t - 3$, and the other is taken from the input at time t, $t - 1$, $t - 2$ and $t - 3$.

$\mathbf{v}^{(1)}$ and $\mathbf{v}^{(2)}$ are obtained as the convolution of the input sequence $\mathbf{u}$. Usually the output sequences are multiplexed into one output sequence.

In the next section we will show how to mathematically show this convolution.

6.2   TIME DOMAIN APPROACH

Let $\mathbf{u} = (10000\ldots)$. We are going to enter this sequence and observe the impulse response from each output. For Figure 4.3 we will get two outputs.

We will obtain length $\nu = (m + 1)$ vectors which we denote by $\mathbf{g}^{(1)}$ and $\mathbf{g}^{(2)}$.

EXAMPLE 4.59 For Figure 4.3 we therefore get

$$
\begin{aligned}
\mathbf{g}^{(1)} &= (1011) \\
\mathbf{g}^{(2)} &= (1111)
\end{aligned}
\tag{4.150}
$$

We can then write the encoding procedure as

$$
\begin{aligned}
\mathbf{v}^{(1)} &= \mathbf{u} * \mathbf{g}^{(1)} \\
\mathbf{v}^{(2)} &= \mathbf{u} * \mathbf{g}^{(2)}
\end{aligned}
\tag{4.151}
$$

where $*$ represents the convolution operation. More explicitly we get

$$
v_l^{(j)} = \sum_{i=0}^{m} u_{l-i} g_i^{(j)} = u_l g_0^{(j)} + u_{l-1} g_1^{(j)} + \cdots + u_{l-m} g_m^{(j)}
\tag{4.152}
$$

So for the encoder of Figure 4.3 we get

$$
\begin{aligned}
v_l^{(1)} &= u_l + && u_{l-2} + u_{l-3} \\
v_l^{(2)} &= u_l + u_{l-1} + u_{l-2} + u_{l-3}
\end{aligned}
\tag{4.153}
$$

so for $\mathbf{u} = (10111)$ we get

$$
\begin{aligned}
\mathbf{v}^{(1)} &= (10000001) \\
\mathbf{v}^{(2)} &= (11011101)
\end{aligned}
\tag{4.154}
$$

Note here we take $u_l = 0$ for $l < 0$ and $l < length(\mathbf{u})$. If these sequences are then multiplexed in the usual manner we get the output sequence as

$$
\mathbf{v} = (1101000101010011).
\tag{4.155}
$$

6.3   THE GENERATOR MATRIX

In the same way as block codes we can (theoretically anyway) produce a generator matrix for a convolutional code. The word theoretically is used here because as we now convolutional codes act on sequences that are potentially infinite in length and as such the generator may need to be infinite. We will explain more as we go along.

If the generator sequences are for a $(2, 1, m)$ convolutional code, i.e. $\mathbf{g}^{(1)}$ and $\mathbf{g}^{(2)}$, then if we place them in the matrix as follows:

$$
G = \begin{bmatrix}
g_0^{(1)} & g_0^{(2)} & g_1^{(1)} & g_1^{(2)} & \cdots & g_m^{(1)} & g_m^{(1)} & & & \\
& & g_0^{(1)} & g_0^{(2)} & \cdots & g_{m-1}^{(1)} & g_{m-1}^{(1)} & g_m^{(1)} & g_m^{(1)} & \\
& & & \ddots & & & & \ddots & \ddots & \ddots & \ddots
\end{bmatrix}
\tag{4.156}
$$

we can now write

$$
\mathbf{v} = \mathbf{u}G
\tag{4.157}
$$

We say that this matrix is *semi-infinite*, corresponding to the length of $\mathbf{u}$. If the input vector $\mathbf{u}$ is finite of length L, say, then G has L rows and $2(m + L)$ columns, giving $\mathbf{v}$ of length $2(m + L)$.

EXAMPLE 4.60 For the convolutional encoder given in Figure 4.3 we get the following generator matrix.

$$
G = \begin{bmatrix}
1 & 1 & 0 & 1 & 1 & 1 & 1 & 1 & & & \\
& 1 & 1 & 0 & 1 & 1 & 1 & 1 & 1 & & \\
& & 1 & 1 & 0 & 1 & 1 & 1 & 1 & 1 & \\
& & & 1 & 1 & 0 & 1 & 1 & 1 & 1 & 1 \\
& & & & \ddots & & \ddots & & \ddots & & \ddots
\end{bmatrix}
\tag{4.158}
$$

So if for example we have the same input sequence as Example 4.59, then we have $\mathbf{u} = (10111)$ and

$$
G = \begin{bmatrix}
1 & 1 & 0 & 1 & 1 & 1 & 1 & 1 & & & & & & \\
 & 1 & 1 & 0 & 1 & 1 & 1 & 1 & 1 & & & & & \\
 & & 1 & 1 & 0 & 1 & 1 & 1 & 1 & 1 & & & & \\
 & & & 1 & 1 & 0 & 1 & 1 & 1 & 1 & 1 & & & \\
 & & & & 1 & 1 & 0 & 1 & 1 & 1 & 1 & 1 & & \\
 & & & & & 1 & 1 & 0 & 1 & 1 & 1 & 1 & 1 \\
\end{bmatrix}
\tag{4.159}
$$

and after encoding via (4.157) we get

$$
\mathbf{v} = (1101000101010011) \tag{4.160}
$$

which is the same as we produced before in (4.155).

6.4   TRANSFORM DOMAIN APPROACH

Since a convolutional encoder is a linear system, each output sequence can be replaced by a corresponding polynomial.

EXAMPLE 4.61   For the encoder in Figure 4.3 using the transform domain we therefore get

$$
\begin{aligned}
\mathbf{v}^{(1)}(D) &= \mathbf{u}(D)\mathbf{g}^{(1)}(D) \\
\mathbf{v}^{(2)}(D) &= \mathbf{u}(D)\mathbf{g}^{(2)}(D)
\end{aligned}
\tag{4.161}
$$

where $\mathbf{u}(D) = u_0 + u_1 D + \cdots$.

The $\mathbf{g}^{(i)}(D), i = 1, 2$ are known as the *generator polynomials*. As we usually multiplex the output streams into a single stream we need to provide a mathematical way to do this. For our two output stream system we can write

$$
\mathbf{v}(D) = \mathbf{v}^{(1)}(D^2) + D\mathbf{v}^{(1)}(D^2) \tag{4.162}
$$

The indeterminate D can be interpreted as the delay operator.

Since the last stage of the shift register in an $(n, 1, m)$ code must be connected to at least one output

$$
m = \max(\deg[\mathbf{g}^{(i)}(D)] : 0 < j < n - 1). \tag{4.163}
$$

6.5   POLYNOMIAL MATRIX REPRESENTATION

In matrix form we get

$$
\mathbf{V}(D) = \mathbf{U}(D)\mathbf{G}(D) \tag{4.164}
$$

where $\mathbf{U}(D) = [\mathbf{u}^{(1)}(D), \mathbf{u}^{(2)}(D), \ldots, \mathbf{u}^{(k)}(D)]$ is the k-tuple of input sequences, $\mathbf{V}(D) = [\mathbf{v}^{(1)}(D), \mathbf{v}^{(2)}(D), \ldots, \mathbf{v}^{(n)}(D)]$ is the n-tuple of output sequences and

$$\mathbf{G}(D) = \begin{bmatrix} \mathbf{g}_1^{(1)}(D) & \mathbf{g}_1^{(2)}(D) & \cdots & \mathbf{g}_1^{(n)}(D) \\ \mathbf{g}_1^{(1)}(D) & \mathbf{g}_1^{(2)}(D) & \cdots & \mathbf{g}_1^{(n)}(D) \\ \vdots & \vdots & \ddots & \vdots \\ \mathbf{g}_k^{(1)}(D) & \mathbf{g}_k^{(2)}(D) & \cdots & \mathbf{g}_k^{(n)}(D) \end{bmatrix} \qquad (4.165)$$

Here $\mathbf{g}_i^{(j)}(D)$ is interpreted as the generator polynomial for the ith encoder input to the jth output.

Then we can see that

$$\mathbf{v}(D) = \mathbf{v}^{(1)}(D^n) + D\mathbf{v}^{(2)}(D^n) + \cdots + D^{n-1}\mathbf{v}^{(n)}(D^n) \qquad (4.166)$$

$\mathbf{G}$ is also known as the *transfer function matrix*.

EXAMPLE 4.62 Let us again look at the encoder in Figure 4.3 which we display here again as Figure 4.4.

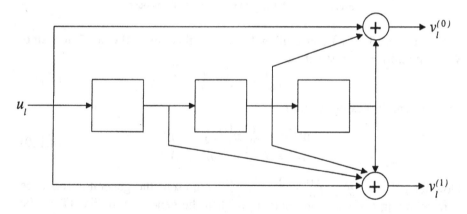

Figure 4.4.   A $(2, 1, 3)$ Convolutional Encoder

We see that the generator matrix is

$$\mathbf{G}(D) = [1 + D^2 + D^3, 1 + D + D^2 + D^3] \qquad (4.167)$$

6.6   THE CASE $K > 1$

We see in Figure 4.5 the case of a convolutional encoder for the case when $k = 2$. As can be seen this is far more complex. The parameters for this encoder are $(3, 2, 1)$.

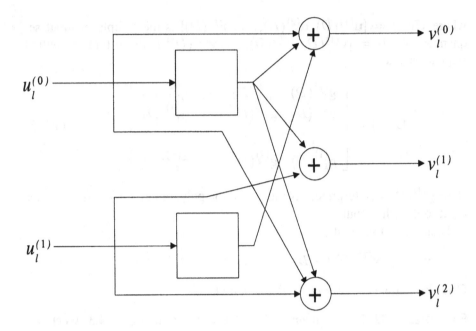

Figure 4.5.   A $(3, 2, 1)$ Convolutional Encoder

Consider a $(3, 2, m)$ convolutional code. In this bits will enter 2 at a time. Write the information stream as

$$\mathbf{u} = (u_0^{(1)} u_0^{(2)} u_1^{(1)} u_1^{(2)} u_2^{(1)} u_2^{(2)} \ldots) \tag{4.168}$$

or after demultiplexing we have

$$\begin{aligned} \mathbf{u}^{(1)} &= (u_0^{(1)} u_1^{(1)} u_2^{(1)} \ldots) \\ \mathbf{u}^{(2)} &= (u_0^{(2)} u_1^{(2)} u_2^{(2)} \ldots) \end{aligned} \tag{4.169}$$

Now let $\mathbf{g}_i^{(j)} = (g_{i,0}^{(j)}, g_{i,1}^{(j)}, g_{i,2}^{(j)}, \ldots, g_{i,m}^{(j)})$ represent the generator sequence corresponding to inout i and output j. Then the generator matrix, $\mathbf{G}$ can be written as

$$\mathbf{G} = \begin{bmatrix} g_{1,0}^{(1)} & g_{1,0}^{(2)} & g_{1,0}^{(3)} & g_{1,1}^{(1)} & g_{1,1}^{(2)} & g_{1,1}^{(3)} & \cdots & g_{1,m}^{(1)} & g_{1,m}^{(2)} & g_{1,m}^{(3)} \\ g_{2,0}^{(1)} & g_{2,0}^{(2)} & g_{2,0}^{(3)} & g_{2,1}^{(1)} & g_{2,1}^{(2)} & g_{2,1}^{(3)} & \cdots & g_{2,m}^{(1)} & g_{2,m}^{(2)} & g_{2,m}^{(3)} \\ & & & g_{1,0}^{(1)} & g_{1,0}^{(2)} & g_{1,0}^{(3)} & \cdots & g_{1,m-1}^{(1)} & g_{1,m-1}^{(2)} & g_{1,m-1}^{(3)} & g_{1,m}^{(1)} & g_{1,m}^{(2)} & g_{1,m}^{(3)} \\ & & & g_{1,0}^{(1)} & g_{1,0}^{(2)} & g_{1,0}^{(3)} & \cdots & g_{1,m-1}^{(1)} & g_{1,m-1}^{(2)} & g_{1,m-1}^{(3)} & g_{1,m}^{(1)} & g_{1,m}^{(2)} & g_{1,m}^{(3)} \\ & & & & \ddots & \ddots & & \ddots & & & & \ddots \end{bmatrix} \tag{4.170}$$

EXAMPLE 4.63 For the encoder of Figure 4.5 we get

$$\begin{aligned} \mathbf{g}_1^{(1)} &= (11), & \mathbf{g}_1^{(2)} &= (01), & \mathbf{g}_1^{(3)} &= (11) \\ \mathbf{g}_2^{(1)} &= (01), & \mathbf{g}_2^{(2)} &= (10), & \mathbf{g}_2^{(3)} &= (10) \end{aligned} \tag{4.171}$$

and the encoding equations can be written

$$
\begin{aligned}
\mathbf{v}^{(1)} &= \mathbf{u}^{(1)} * \mathbf{g}_1^{(1)} + \mathbf{u}^{(2)} * \mathbf{g}_2^{(1)} \\
\mathbf{v}^{(2)} &= \mathbf{u}^{(1)} * \mathbf{g}_1^{(2)} + \mathbf{u}^{(2)} * \mathbf{g}_2^{(2)} \\
\mathbf{v}^{(3)} &= \mathbf{u}^{(1)} * \mathbf{g}_1^{(3)} + \mathbf{u}^{(2)} * \mathbf{g}_2^{(3)}
\end{aligned}
\tag{4.172}
$$

Then the convolution operation gives

$$
\begin{aligned}
v_l^{(1)} &= u_l^{(1)} & &+ u_{l-1}^{(1)} &+ u_{l-1}^{(2)} \\
v_l^{(2)} &= & u_l^{(2)} &+ u_{l-1}^{(1)} & \\
v_l^{(3)} &= u_l^{(1)} &+ u_l^{(2)} &+ u_{l-1}^{(1)}, &
\end{aligned}
\tag{4.173}
$$

after multiplexing we get

$$
\mathbf{v} = (v_0^{(1)} v_0^{(2)} v_0^{(3)}, v_1^{(1)} v_1^{(2)} v_1^{(3)}, v_2^{(1)} v_2^{(2)} v_2^{(3)}, \ldots)
\tag{4.174}
$$

Take

$$
\mathbf{u} = (110110),
\tag{4.175}
$$

so

$$
\begin{aligned}
\mathbf{u}^{(1)} &= (101) \\
\mathbf{u}^{(2)} &= (110)
\end{aligned}
\tag{4.176}
$$

Then we can use either the convolutional method as we have seen, or if we use the generator matrix

$$
G =
\begin{bmatrix}
1 & 0 & 1 & 1 & 1 & 1 & & & & & & \\
0 & 1 & 1 & 1 & 0 & 0 & & & & & & \\
& & 1 & 0 & 1 & 1 & 1 & 1 & & & & \\
& & 0 & 1 & 1 & 1 & 0 & 0 & & & & \\
& & & & 1 & 0 & 1 & 1 & 1 & 1 & & \\
& & & & 0 & 1 & 1 & 1 & 0 & 0 &
\end{bmatrix}
\tag{4.177}
$$

then $\mathbf{v} = \mathbf{u}G$ gives:

$$
\mathbf{v} = (110, 000, 001, 111).
\tag{4.178}
$$

6.7   SYSTEMATIC AND NON-SYSTEMATIC CONVOLUTIONAL CODES

Consider the encoder of Figure 4.6. This is a $(2, 1, 2)$ convolutional code.

It is easy to see that for this encoder there does not exists an exact replication of the input stream in the output stream. This convolutional code is therefore *non-systematic*. The polynomial generator matrix of this code is easily seen to be

$$
\mathbf{G}(D) = [\, 1 + D^2 \quad 1 + D + D^2 \,].
\tag{4.179}
$$

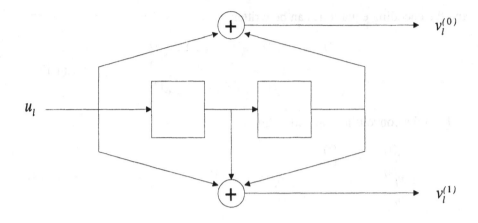

Figure 4.6.   A $(2, 1, 2)$ Convolutional Encoder

However now consider the convolutional code given in Figure 4.7. This is once again a $(2, 1, 2)$ convolutional code, but this time it is *systematic*, as we can see clearly that there is an exact replication of the input stream occurring in the output stream.

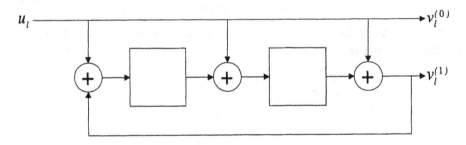

Figure 4.7.   A $(2, 1, 2)$ Systematic Convolutional Encoder

This encoder in fact produces an equivalent code to the encoder of Figure 4.6. Equivalence here is determined up to permutation of the code bits.

The polynomial generator matrix for the encoder of Figure 4.7 is

$$\mathbf{G}(D) = [\; 1 \quad (1 + D + D^2)/(1 + D^2) \;]. \qquad (4.180)$$

which is (a simple) systematic generator matrix as the identity exists at the left of $\mathbf{G}$.

Note here that the encoder given in Figure 4.7 is also recursive in that bit actually going into the encoder is also determined by a feedback loop from the output. This kind of encoder is known Recursive Systematic Convolutional Encoder . Of course it is possible to have Non-recursive Systematic Convolutional Encoders and Recursive Non-sysematic Convolutional Encoders and these are all obvious variations on the Figures 4.7 and 4.6.

6.8   DISTANCE PROPERTIES

As in block codes, the performance of convolutional codes depends strongly on the distance properties of the code. However this is not such an easy task as we do not have code words as such, because of the potentially infinite size of the input and output words.

For convolutional codes we will define a property known as the *free distance*.

DEFINITION 4.64 The *free distance*, d_{free} is defined as

$$
\begin{aligned}
d_{free} &= \min[d(\mathbf{v}^{(0)}, \mathbf{v}^{(1)}) : \mathbf{u}^{(0)} \neq \mathbf{u}^{(0)}] \\
&= \min[\text{wt}(\mathbf{u}\mathbf{G}) : \mathbf{u} \neq \mathbf{0}],
\end{aligned}
\tag{4.181}
$$

where $\mathbf{v}^{(i)}, i = 0, 1$ is the code word corresponding to the information vector $\mathbf{u}^{(i)}, i = 0, 1$, $d(\mathbf{v}^{(0)}, \mathbf{v}^{(1)})$ is the (Hamming) distance between $\mathbf{v}^{(0)}$ and $\mathbf{v}^{(1)}$, respectively, and wt($\mathbf{x}$ is the (Hamming) weight of the vector $\mathbf{x}$. Once again we reiterate that all the vectors involved can be infinite in length.

In general d_{free} has not been found, but for specific codes there are some interesting and useful results.

6.9   GRAPHICAL REPRESENTATION
TREE DIAGRAM

Consider again the encoder given in Figure 4.6. Consider the encoder to be starting with the memory (the state) set to zero. Then we can represent the code as in Figure 4.8.

The diagram is representing the input bit (a 0 is up and a 1 is down) and the output pair of bits (which are labelled on the branches). The main problem with this is that it gets exponentially larger so is not a realistic way to look at the output stream. However if we look closely we can see that there is repetition. After the third branch we can see a replication of the previous three branches. Look at Figure 4.9, and looking at nodes 1 and 1' we see an identical tree growing to the right.

We can therefore merge the nodes 1 and 1' (and similarly 2 with 2', 3 with 3' and 4 with 4', to give a new way of looking at the transitions which we will look at more closely in the next section.

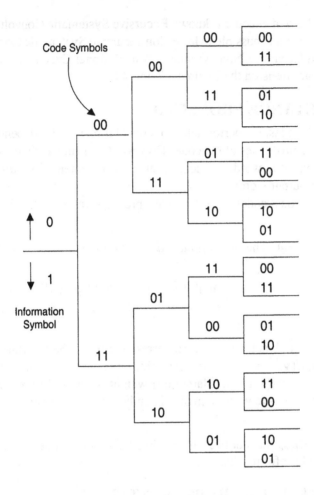

Figure 4.8.   Tree diagram of Figure 4.6

TRELLIS DIAGRAM

The merging that was explained in the previous section would result in the object known as a *trellis*. For the tree explained in the previous section the corresponding trellis can be seen in Figure 4.10. A dotted line represents an input of 1 and a solid line represents and input of 0.

We will develop the ideas of trellises later and indeed further we will use the trellis as a means of decoding the convolutional code. First of all we will look at the state diagram of a convolutional code.

6.10   ENCODER STATE DIAGRAM

Since the state of the encoder is defined as the contents of the shift register elements, the entire operation can be described by a *state diagram*

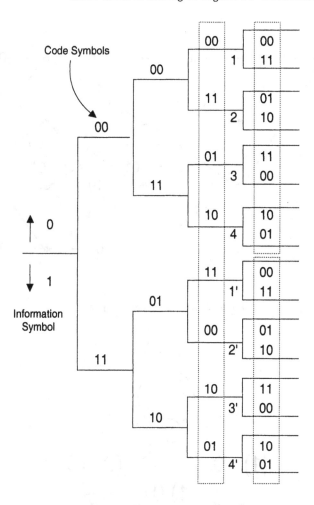

Figure 4.9.   Tree diagram showing repetitions

Consider again the convolutional code of Figure 4.6. We get the state diagram given in Figure 4.11.

It is easy to see the relationship between the state diagram and the trellis diagram. Of course the state diagram has no concept of time and this is the extra axis that is present in the trellis. For encoding though either method could be used.

6.11   DECODING OF CONVOLUTIONAL CODES

Consider an (n, k, m) convolutional code, with constraint length, ν. Let the information word be

$$\mathbf{u} = (\mathbf{u}_0, \mathbf{u}_1, \ldots, \mathbf{u}_{L-1}). \tag{4.182}$$

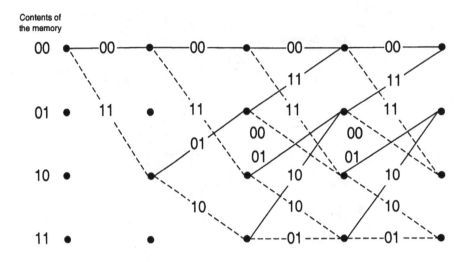

Figure 4.10.   Trellis of Figure 4.6

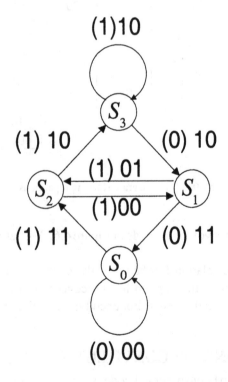

Figure 4.11.   State diagrma of Figure 4.6

We will consider only the concept of a finite length information word. The length of this vector is kL in total. Consider that this is encoded to

$$\mathbf{c} = (\mathbf{c}_0, \mathbf{c}_1, \ldots, \mathbf{c}_{N-1}). \tag{4.183}$$

This encoded word has total length nkL. $\mathbf{c}$ is then sent over a discrete memoryless channel, and the following is received.

$$\mathbf{r} = (\mathbf{r}_0, \mathbf{r}_1, \ldots, \mathbf{r}_{N-1}). \tag{4.184}$$

We need to find a decoder such that given $\mathbf{r}$ we want to find an estimate $\hat{\mathbf{c}}$ of the original codeword $\mathbf{c}$.

6.12   OPTIMAL DECODING

We need to look at the probabilities associated with decoding. The probability that we received $\mathbf{r}$ and decode it in error is

$$P(E|\mathbf{r}) = P(\hat{\mathbf{c}} \neq \mathbf{c}|\mathbf{r}). \tag{4.185}$$

So the error probability of the decoder is

$$P(E) = \sum_{\forall \mathbf{r}} P(E|\mathbf{r})P(\mathbf{r}). \tag{4.186}$$

Optimum decoding must minimise $P(E|\mathbf{r}) = P(\hat{\mathbf{c}} \neq \mathbf{c}|\mathbf{r})$ for all $\mathbf{r}$. Alternatively, we want to maximise $P(\hat{\mathbf{c}} = \mathbf{c}|\mathbf{r})$. $P(E|\mathbf{r})$ is minimised for a given $\mathbf{r}$ by choosing $\hat{\mathbf{c}}$ as the codeword $\mathbf{c}$ which maximises

$$P(\mathbf{c}|\mathbf{r}) = \frac{P(\mathbf{r}|\mathbf{c})P(\mathbf{v})}{P(\mathbf{r})}. \tag{4.187}$$

If all information sequences are equally likely, then $P(\mathbf{c})$ is the same for all $\mathbf{c}$ and maximising $P(\mathbf{c}|\mathbf{r})$ is equivalent to maximising $P(\mathbf{r}|\mathbf{c})$.

Since for a DMC (i.e. each bit is statistically independend from the other bits of the word)

$$\begin{aligned} P(\mathbf{r}|\mathbf{c}) &= \prod_{i=0}^{L-1} P(\mathbf{r}_i|\mathbf{c}_i) \\ &= \prod_{j=0}^{N-1} P(r_j|c_j) \end{aligned} \tag{4.188}$$

it follows that

$$\begin{aligned} \log[P(\mathbf{r}|\mathbf{c})] &= \sum_{i=0}^{L-1} \log[P(\mathbf{r}_i|\mathbf{c}_i)] \\ &= \sum_{j=0}^{N-1} \log[P(r_j|c_j)] \end{aligned} \tag{4.189}$$

where $\log[P(r_j|c_j)]$ is the channel transition probability.

6.13   MAXIMUM LIKELIHOOD DECODING

Further to the decoding we have been looking at for block codes we would like to find a *maximum likelihood decoder*(MLD). An MLD chooses ĉ as the codeword c which maximises the log likelihood ratio function $\log[P(\mathbf{r}|\mathbf{c})]$. We can see in Figure 4.12, the relationship between an MLD for a block code and an MLD for a convolutional code.

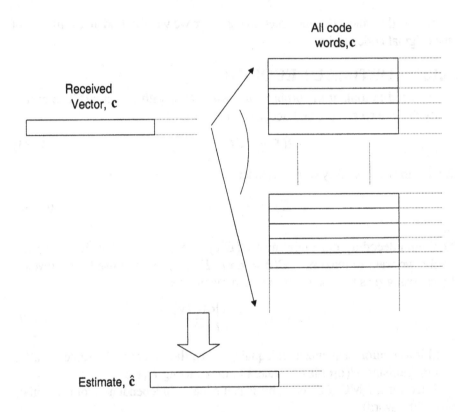

Figure 4.12.   Maximum Likelihood Decoding

If we look closely we see that for a block code (the solid lines) there are a finite set of codewords so in theory (depending on the size of the code of course) we can develop a decoder that looks at all the codewords.

However for convolutional code we have a potentially infinite number of codewords (which are potentially infinite in length) and as such we can never look at all of them.

6.14   PRACTICAL ENCODERS

In order for a decoder to perform as expected we need the information word, **u** and the codeword, **c** to be one-to-one and reversible. Further in the polynomial form we need $u(D)$ and $v(D)$ to also be one-to-one and reversible. Thus we neeed a $G^{-1}(D)$ such that

$$G(D)G^{-1} = ID^N \qquad (4.190)$$

where I is the $(k \times k)$ identity matrix, for some values N. Now if we can for the moment consider that the information words are of finite length then we can do the following:

$$\begin{aligned} c(D)G(D) &= u(D)G^{-1}G(D) \\ &= u(D)D^N \end{aligned} \qquad (4.191)$$

and the information word can be recovered with delay N. Of course we can theoretically do the above calculations with semi-infinite matrices.

If there does not exist a $G^{-1}(D)$ then the encoder is known as *catastrophic*, a property that we will investigate further in the next section.

6.15   CATASTROPHIC ENCODERS

DEFINITION 4.65 A convolutional code defined by the polynomial generator matrix $G(D)$ is said to be *catastrophic* if there is an infinite weight vector $u(D)$ such that the corresponding codeword $c(D) = u(D)G(D)$ has finite weight.

We can interpret this as a finite number of channel errors can result in an infinite number of decoding errors.

THEOREM 4.66 *For an $(n, 1, m)$ convolutional code, if*

$$GCD[g^{(1)}(D), g^{(2)}(D), \ldots, g^{(n)}(D)] = D^N \qquad (4.192)$$

for some N, then the convolutional code is non-catastrophic.

COROLLARY 4.67 *All systematic codes are non-catastrophic.*

Consider the encoder given in Figure 4.13. This is very similar to the encoder given before in Figure 4.6. However we will see there are serious problems with this one, in that it is a catastrophic encoder.

Let the information word be $u = (11111\ldots)$. This is encoded via the encoder in Figure 4.13 to $c = (1110000\ldots)$.

Now if three errors occur in the transmission (in the first three places) then we would receive $r = (000000\ldots)$. Using an MLD decoder we would decode this to $\hat{u} = (00000\ldots)$.

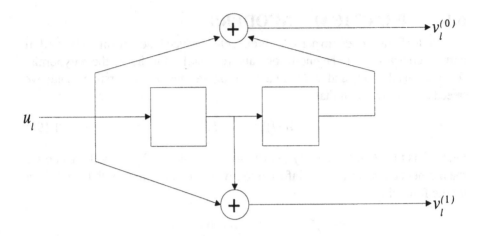

Figure 4.13.   A $(2, 1, 2)$ Catastrophic Convolutional Encoder

Therefore we can see that a finite number of channel errors causes an infinite number of decoding errors.

To see the problem here we need to look at the state diagram of the encoder in Figure 4.13. This is shown in Figure 4.14.

The main point to note about the state diagram in Figure 4.14 is that there is a loop of zero weight (at S_3) other then the self loop around the state S_0. This indicates a catastrophic encoder.

6.16   STATE DIAGRAMS WHEN $K > 1$

We have seen in Section 6.6 that the encoder is more complex for $k > 1$. In fact for the encoder given in Figure 4.7, which we display here again as Figure 4.15.

We can see that for each clock tick input there will be two input bits. This then indicates that we can have $4 = 2^2$ different edges on the state diagram, and indeed we get Figure 4.16 as the state diagram for the encoder in Figure 4.15. This is a $(3, 2, 1)$ convolutional code. Note that there are four inputs and four outputs at each node on the state diagram.

7.   VITERBI DECODING

In [208] Viterbi showed that there was a way to traverse the trellis such that the result of the traverse was a decoding that was optimal in the sense of maximum likelihood.

Recall that the log likelihood function $\log[P(\mathbf{r}|\mathbf{c})]$ and define it as the *metric* associated with the path, $\mathbf{c}$, and denote by $M(\mathbf{r}|\mathbf{c})$. Also define $\log[P(\mathbf{r}_i|\mathbf{c}_i)]$ as the *branch metrics* and denote by $M(\mathbf{r}_i|\mathbf{c}_i)$. Further define $\log[P(r_i|c_i)]$ by

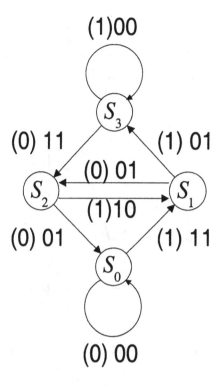

Figure 4.14.   State diagram of Figure 4.13

the *bit metrics* and denote by $M(r_i|c_i)$. Hence

$$
\begin{aligned}
M(\mathbf{r}|\mathbf{c}) &= \sum_{i=0}^{L-1} M(\mathbf{r}_i|\mathbf{c}_i) \\
&= \sum_{i=0}^{N-1} M(r_i|c_i)
\end{aligned} \tag{4.193}
$$

where we are looking at received word of length L symbols, or $N = kL$ bits. A partial metric for the first j branches of a path can be expressed as

$$
M(\mathbf{r}|\mathbf{c}) = \sum_{i=0}^{j-1} M(\mathbf{r}_i|\mathbf{c}_i). \tag{4.194}
$$

We are now in a position to state the Viterbi algorithm.

1. Beginning at time $j = m$, compute the partial metric for the single path entering each state

2. Increase j by 1. Compute the partial metric for all paths entering by adding the branch metric entering the state to the metric of the connecting path

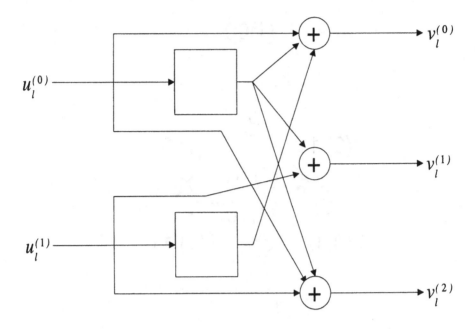

Figure 4.15.   A $(3, 2, 1)$ Convolutional Encoder

from the preceding time unit. For each state, store the path with the largest metric, together with its metric (the *survivor*) eliminate all other paths

3. If $j < L + m$, repeat step 2. Otherwise stop.

Of course the power of this algorithm is two-fold, firstly it has been proven to be a maximum likelihood decoder and secondly that it can be used on a infinite sequence. In order for the latter to occur we need to rescale the metrics when necessary but this is possible as we all the calculations are simply summations so we can just rescale the smallest to zero, then we will never get an overflow error.

Let us now briefly explain what is known as the *sliding window method*. This is such that if we have a set window length, l, and once we reach time $t = l$ we simple output the best estimate bit at time $t = 0$, then at $t = l + 1$ we output the best estimate bit at time $t = 1$ and so on. This means that we never need to have more than the window size of surviving paths in the memory.

THEOREM 4.68 *When using the sliding window method, if the window length is set to 6ν then with high probability the output bit will be such that all the surviving paths will have converged to it. This is not a sufficient condition as the Example 4.69 will show.*

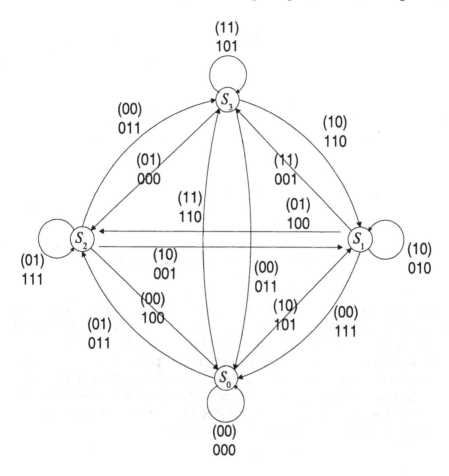

Figure 4.16.   State diagram of Figure 4.15

The metrics we use can vary depending on the complexity aspects, the channel and indeed the implementor. In the Example 4.69 we will use the Hamming distance as the metric. So we will need to minimise the accumulated metric. This method of decoding is known as *hard decision Viterbi decoding* . The probabilistic method is known as *soft decision Viterbi decoding* . The Example 4.69 we will highlight some interesting issues that can come around when we are using the Viterbi

EXAMPLE 4.69 Consider again the code given in Figure 4.6, which we give again here as Figure 4.17.

We will assume that the encoder is in the all zero state prior to encoding.

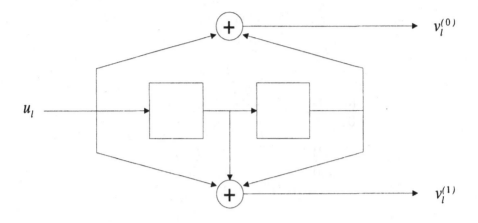

Figure 4.17.   A $(2, 1, 2)$ Convolutional Encoder

If we have the information vector $\mathbf{u} = (1001000\ldots)$, then this is encoded to $\mathbf{v} = (11011111011100\ldots)$. Consider that this is sent across a channel and we demodulate to give the received vector as $\mathbf{r} = (01011111011100\ldots)$.

We know that an error has occurred in the first position. Can we use the Viterbi decoding algorithm to decode this correctly? We will use a window of length 2ν to show that Theorem 4.68 needs to be taken into consideration.

We will now proceed with the Viterbi algorithm for our received word.

The trellis for this code is seen in Figure 4.18.

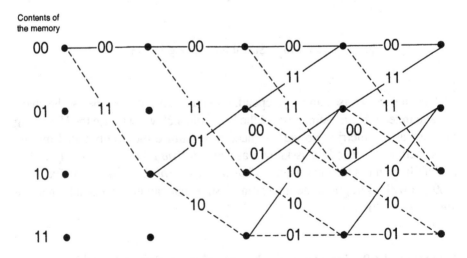

Figure 4.18.   Trellis for Figure 4.17

We can now form the trellis one time step at a time. On the first step we are starting from the all zero state so we only have two possible paths corresponding to an input of 0 or 1. The Hamming distance between the received pair (10) and the options are placed on the trellis. This can be seen in Figure 4.19.

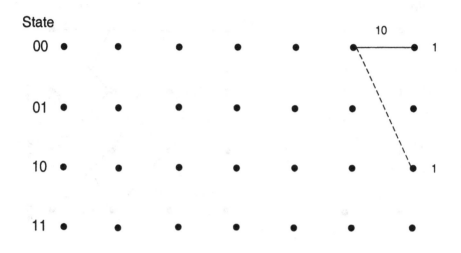

Figure 4.19.

We now move to the next time step. Still we are building up the trellis so there are no survivor paths to find. We see the second time step completed in Figure 4.20.

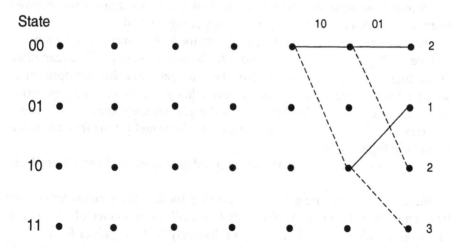

Figure 4.20.

Moving to time step 3 we see we have merging paths, and so it is at this stage that we need to remove some paths. With all the paths on we get Figure 4.21. All the metrics are displayed.

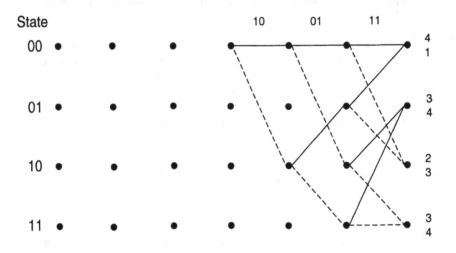

Figure 4.21.

As the Viterbi algorithm states we must look at the metrics and eliminate the paths with the largest (as we are considering the Hamming distance) metric at each node. We thus see in Figure 4.22 the eliminated paths (with the crosses on). We therefore get Figure 4.23 which shows only the surviving paths.

We now move to the next time step, and look at the pair of bits in the received word, i.e. 11. Accumlating the metrics we get Figure 4.24.

Then we need to eliminate the paths with the highest metrics at each node, and we get Figure 4.25. We now move to the next time step. We accumulate the metrics to obtain Figure 4.26. Note here we get into a difficult position as we have to eliminate a path at node 11 even though the two converging pathjs have identical metrics. For this position the general method is to eliminate a arbitrary one. We choose the upper one to eliminate and get the four surviving paths as in Figure 4.27.

We now continue for the next time step and get Figure 4.28 and Figure 4.29.

We can now see we are getting into another dilema. The window length has been set to $2\nu = 6$ and this is obviously too small for the choice of first output bit to be unambiguous. We have two (obviously!!) two choices for the first output bit. The only choice we have is to look at the lowest metric and trace it back to the first time step. This will result in a 1 being output. So we have the first bit output from the decoder.

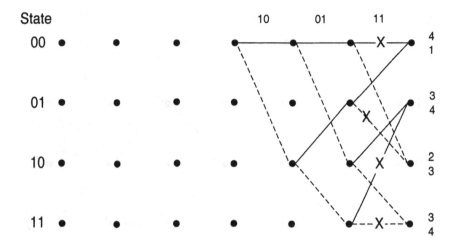

Figure 4.22.

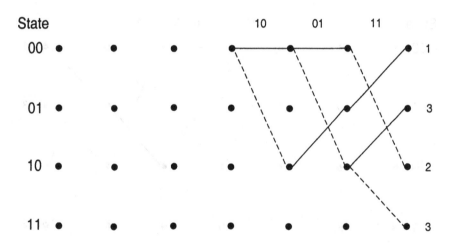

Figure 4.23.

Next we move to the next time step, and get Figure 4.30 and Figure 4.31.

Once again we see that we are in an slightly ambiguous situation in that there is no clear decision for the output bit. We need to look at the metrics and once again we decide to output 0. We can go on with this now and retrieve the information word 100100..., as expected and we have thus corrected one error.

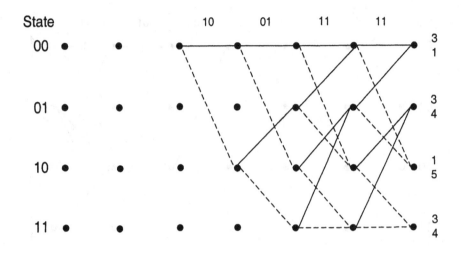

Figure 4.24.

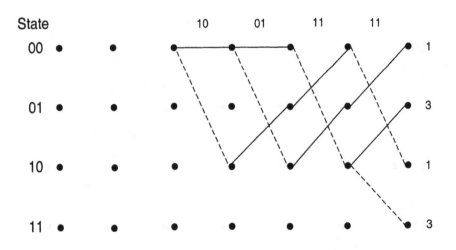

Figure 4.25.

The ambiguity occurring in Example 4.69 is because we are using the hard decision decoding method. There is less chance (and far more accuracy) if we use soft decision decoding (if that is possible).

Indeed if we were to decode the word using a larger window we still would not obtain a definite solution. It is easy to see though that if we also use the metrics then the solution can be obtained as we have already done.

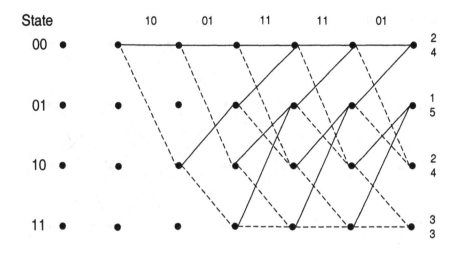

Figure 4.26.

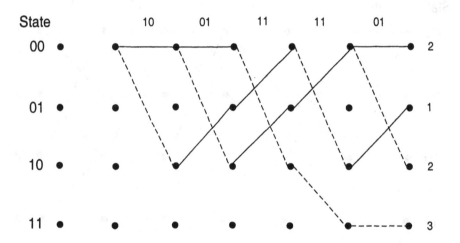

Figure 4.27.

There is of course the issue of termination and what do we do if we actually want to stop the decoder. What happens to the bits in the decoder that have yet to be output. There are two basic ideas for this solution.

1. take the lowest accumulated metric (as we did in Example 4.69, or

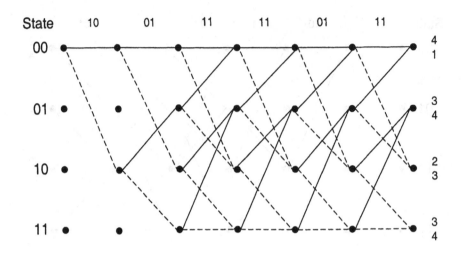

Figure 4.28.

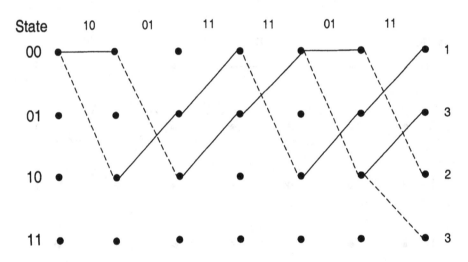

Figure 4.29.

2. encode a *tail* to the information word, that is known to the decoder. The idea of this is to reset the decoder to a known state. Obviously this will reduce the overall rate of the coding scheme.

7.1   SOFT DECISION VITERBI DECODING

As we have explained in the previous section we could take the quantisation to a higher level than one bit (i.e. a 1 or a 0) and introduce an element of

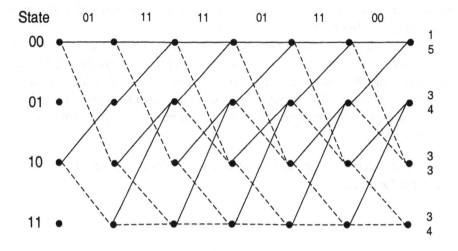

Figure 4.30.

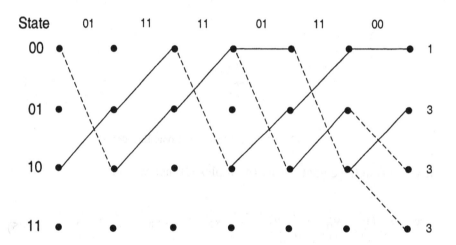

Figure 4.31.

reliability. This then can be applied to the Viterbi algorithm and we can expect gains from hard decision to soft decision of up to 2dB at a BER of 2×10^{-4}.

8. PUNCTURED CONVOLUTIONAL CODES

In this section we will describe a subcalss of (n, k, m) convolutional codes [209, 210, 211]. For simplicity we will be mainly considering an $(n, 1, m)$

binary convolutional code and forming a rate $\frac{p}{q}$ convolutional code. The generalisation to using (n, k, m) codes is easy.

We can obtain a rate $\frac{p}{q}$ (punctured) convolutional code from a $(n, 1, m)$ code by deleting $np - q$ code symbols (bits) from every np code symbols (bits). We are saying bits here to indicate that we are in fact mainly considering binary convolutional codes. The original rate $\frac{1}{n}$ code is known as the *mother code*.

Let us explain this by example. The generalisation to other convolutional codes is obvious.

EXAMPLE 4.70 We once again look at the encoder of Figure 4.17 which we show here as Figure 4.32.

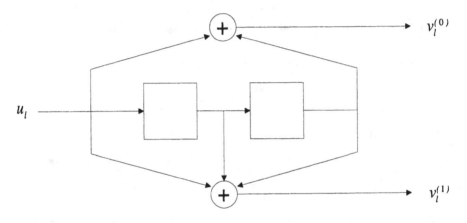

Figure 4.32.   A $(2, 1, 2)$ Convolutional Encoder

If we look at the output stream (multiplexed) then we get

$$\mathbf{v} = (v_0^{(0)}, v_0^{(1)}, v_1^{(0)}, v_1^{(1)}, v_2^{(0)}, v_2^{(1)}, v_3^{(0)}, v_3^{(1)}, v_4^{(0)}, v_4^{(1)}, \\ v_5^{(0)}, v_5^{(1)}, v_6^{(0)}, v_6^{(1)}, \ldots) \qquad (4.195)$$

This is a rate $\frac{1}{2}$ mother code. Let us say we want to form a rate $\frac{2}{3}$ code from this. To do this we simply delete (puncture) every third bit out of four from the stream to get

$$\mathbf{v} = (v_0^{(0)}, v_0^{(1)}, v_1^{(1)}, v_2^{(0)}, v_2^{(1)}, v_3^{(1)}, v_4^{(0)}, v_4^{(1)}, v_5^{(1)}, v_6^{(0)}, v_6^{(1)}, \ldots) \quad (4.196)$$

Obviously the encoding side of the code is only half the problem and so we must address the issue of decoding. It turns out that this is in fact easy, and further we are able to use the Viterbi decoding algorithm. For the punctured positions we simply do not calculate the metrics. For Example 4.70 we see the trellis in Figure 4.33. Where a puncture has occurred an X has been placed. So when we work out the metrics with the Viterbi algorithm we simply miss out the calculation for the bit corresponding to the punctured bits. The puncturing effect can be seen in Figure 4.33.

Contents of
the memory

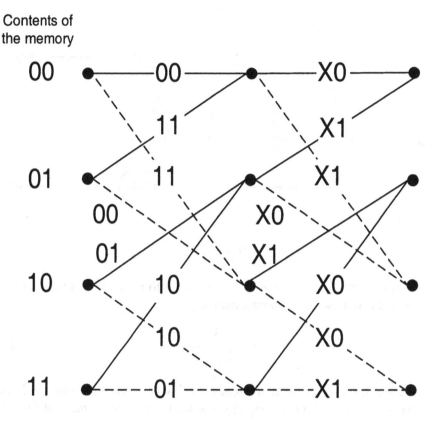

Figure 4.33. Punctured effect on the trellis

It is usual to give the pucturing pattern in terms of a matrix. This will work on the pre-multiplexed pattern, so the *puncturing array* is

$$P = \begin{bmatrix} 1 & 0 \\ 1 & 1 \end{bmatrix} \qquad (4.197)$$

Let us consider a more complex problem. If we look again at the $(3, 2, 1)$ encoder given in Figure 4.5, which we display again here as Figure 4.34.

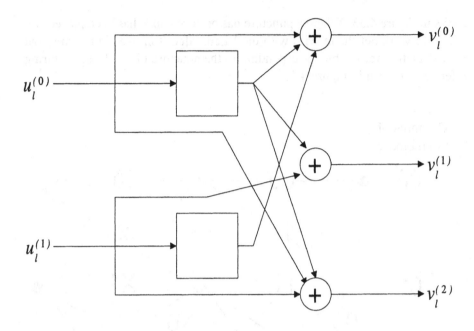

Figure 4.34.

This is a rate $\frac{2}{3}$ mother code. We are going to puncture this to obtain a $\frac{3}{4}$ code. We could use the following matrix:

$$P = \begin{bmatrix} 1 & 0 & 1 \\ 1 & 0 & 0 \\ 0 & 1 & 0 \end{bmatrix} \qquad (4.198)$$

Of course at this stage we don't actually know how good the punctured code is, but we do know that decoding this is at least as easy as decoding the mother code, so implementation is not a problem.

9.   INTERLEAVING AND DEINTERLEAVING

The idea of interleaving stems from the fact that one might get a burst of errors that the decoder cannot handle. That is, for example, say that a $(3, 1, 3)$ repetition code was used. We want to send the two information bits 01. These are encoded to two codewords 000 and 111. However at the receiver we see 011 and 111, an error occurs in position 2 and 3 of the first codeword. We would incorrectly decode these to 11.

However, if we can assume that we start with the two codewords 000 and 111, and consider them as one 6-bit word, i.e. 000111. Before sending the codewords across the channel we shuffle them using the permutation (abcdef) -> (adbecf). We thus get 010101. If once again there were two errors occurring in position

2 and 3 we would receive 001101. We now de-shuffle this to get 010011. Now we split the 6-bit word back to two codewords, 010 and 011. We can now correctly decode this to 01. Of course this involved the window of interest to be larger than one codeword, introducing possible delay and complexity to the system, but we do increase the performance of the system.

There are essentially two types of interleavers. These will be explained in the next two sections.

9.1   BLOCK INTERLEAVERS

The example shown in the first part of the section was a very simple example of a block interleaver. That is we have a window of interest, and the permutation is done only in that window. For the next window we move along the stream the equivalent size of one window. In the example the window was of size 6 bits. So we take 6 bits and work on those six bits only. Then when those 6 bits have been sent we work on the next six bits. Pictorially this can be seen as taking a $(n \times n)$ bits and arranging them in a two dimensional array as in Figure 1.

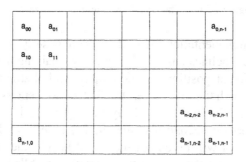

Figure 4.35.   A simple block interleaver

We fill up the array with the $(n \times n)$ bits row wise, i.e. bit 0 goes in $a_{0,0}$, bit 1 to $a_{0,1}$, etc. down to bit $(n - 1) \times (n - 1)$ that goes into position $a_{n-1,n-1}$. Then the bits are read out in a column wise fashion, i.e. bit 0 is $a_{0,n-1}$, bit $(n - 1)$ is $a_{n,n}$, bit $(n - 1) \times (n - 2)$ is $a_{0,0}$ and bit $(n - 1) \times (n - 1)$ is $a_{n-1,n}$. This method is sometimes known as rectangular interleaving. Other block interleaver are based on this in that you fill up an array, but the removal and transmission of the bits is done is other ways. For example a (pseudo) random block interleaver will remove the bits in a pseudo random fashion based on an algorithm for selecting the bits. Of course with this method once a bit has been removed (and sent) it cannot be removed again. In other words, as expected, the rate of the interleaver is 1.

9.2 CONVOLUTIONAL INTERLEAVERS

In the DVB standards there is another type of interleaver that works in a completely different way than block interleavers. These are known as convolutional interleavers, and are based on the interleavers proposed in [194] and [195].

The convolutional interleaver and deinterleaver as used in the DVB standards can be seen in Figure 4.36.

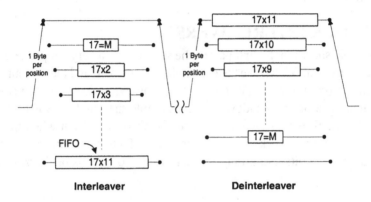

Figure 4.36.   The DVB Convolutional Interleaver

This particular implementation is based on a byte wise system, not a bit wise system. Each of the pipelines has a differing delay, enabling the input symbols to be placed at different positions in the output symbol stream. Concerning ourselves primarily with the interleaver, the first symbol enters the system and and goes along a pipeline with zero delay (i.e. goes straight through). The second symbol enters the second pipeline and has a delay of 17 symbols. So before this one symbol will be output there will be 17 other symbols before it. At the initial stage of the system the pipelines are filled up with zeroes, which will appear in the stream being sent. These zeros will be removed at the deinterleaver. The filling up continues, so that if the input stream was as in Figure 4.37 then we get the output of the interleaver as Figure 4.38. To clarify the numbering of the output symbols, input symbol S_{204} occurs at output position 204 and S_1 occurs at position 205. The deinterleaver has the reverse effect on the stream and so we will get out $(17 \times 11 \times 12)$ zeros and then the input stream as in Figure 4.37.

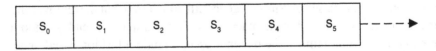

Figure 4.37.   The input stream

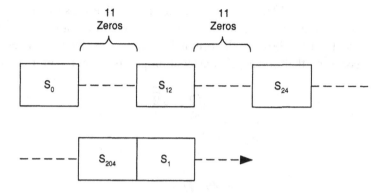

Figure 4.38. The interleaved stream

So it can be seen that the delay introduced by this scheme occurs only at the very beginning of the transmission. As such once the system has been turned on and locked on to, there is no delay, and no deletions are necessary.

10. TRELLIS CODED MODULATION

Traditional channel coding schemes are based on encoding and modulation being performed separately. In the same way in the decoder part of the scheme the demodulation and decoding are performed separatley as well. The error control part of the scheme is provided by means of adding redundancy to the information stream. This has the effect of lowering the information bit rate per channel bandwidth. So as Shannon [200] states bandwidth efficiency is traded off against increased power effeciency.

If we consider that a loss of reliability can occur while passing in information between the modulator and decoder then we would ideally like to join the two together into a modulation coding scheme. This will have better utilisation of the available bandwidth and power. To define this process we want to impose certain patterns on the transmitted signal.

This joint attack at the problem is known as *trellis coded modulation*. So for the same data rate we can use a modulation scheme with an increased number of points. These additional points will allow for redundancy to exists for the same data rate.

The first occurance of this type of scheme was investigated by Ungerboeck [218, 219].

The idea is to use the Euclidean distance instead of the Hamming distance when calculating the best estimate codeword. With the technique we are able to partition an M-ary constellation into $2, 4, 8, \ldots$ subsets each with $M/2, M/4, M/8, \ldots$ points and having increasingly larger Euclidean distance between the points in the constellation.

If we look at Figure 4.39 then we can see the consecutive subsets. In this we have $d_0 < d_1 < d_2$ and so we can have less error protection on the four final subsets than on the two above or the original constellation. The distances are as follows, assuming that the 8PSK constellation points sit on the unit circle.

$$\begin{aligned} d_0 &= \sqrt{2 - \sqrt{2}} \\ d_1 &= \sqrt{2} \\ d_2 &= 2 \end{aligned} \qquad (4.199)$$

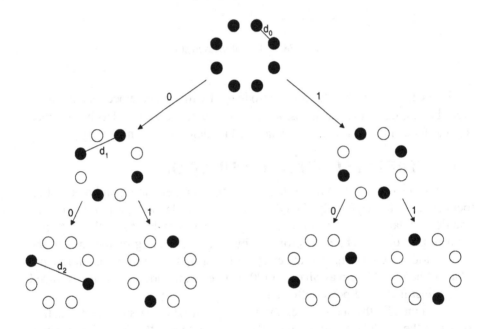

Figure 4.39.   TCM for 8PSK

As an example consider the encoder of Figure 4.40. This is for encoding a bit stream and using 8PSK modulation. The power of it comes from the fact that even though we have an uncoded branch, the Euclidean distance for this is much larger than the Euclidean distance for the whole 8PSK constellation.

Here we have the bit stream demultiplexed to two bit streams. Then the bits enter the system one bit at a time. One of the bits is encoded via a rate $\frac{1}{2}$ code to give two coded bits. Then the bits are mapped to the 8PSK signal constellation points.

For decoding we would use the decoder to work out which of the four final subsets we are at, then as $d_2 > d_1$ and $d_2 > d_0$ we can use the power of the code to offset the fact that we have a smaller Euclidean distance. The maximum Euclidean distance subset then decides the uncoded bit.

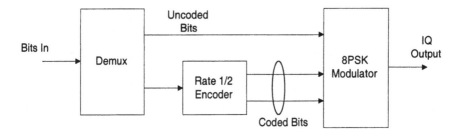

Figure 4.40. A simple TCM encoder

The overall rate of this code is easily seen to be $\frac{2}{3}$, and the scheme is known as rate $\frac{2}{3}$ 8PSK TCM .

This scheme is easily generalised to any rate, by using other encoders. For example if we were to use a rate $\frac{2}{3}$ code and the demultiplexor splits the input stream to three streams then we would get four bits at a time entering the modulator, obviously enabling us to use 16QAM. This is then a rate $\frac{3}{4}$ 16QAM TCM scheme.

It must be noted here that the mapping of the n-tuples to the constellation points is generally different than the usual mappings that we have seen in the previous chapter. These mappings depend on the actual code used and also on the exact method of dividing the subsets. The reader is refered to many of the books on modulation for further reading.

11. PRAGMATIC TRELLIS CODED MODULATION

In the paper [220] it was envisaged that the idea of using TCM schemes for bandwidth and power-limited channels could be generalised by using a single basic code. We have seen in the previous section that we can develop TCM schemes of almost any rate, but for each we can potentially have a different constituent encoder. Therefore in the decoder/demodulator we will have different schemes. In the paper [220], the idea was to use the same encoder (and therefore decoder) and so using the same off the shelf device we are now able to achieve many different rates. The method of this was to allow for various portions of the input to be uncoded and also to allow for puncturing in the encoded streams.

The performance of these generalised schemes is inferior to the best possible schemes using pure TCM, but this is outweighed by the fact that we can more easily implement the system as the constituent parts all exist.

Some versions of this scheme were chosen for the DVB-DSNG standard [221], and will be exemplified in the next chapter.

Chapter 5

EXISTING STANDARDS FOR DIGITAL TV BROADCASTING

1. MAJOR DVB STANDARDS

1.1 DVB-S STANDARD

The standard for digital satellite broadcasting (DVB-S) is one of the oldest standards accepted by the DVB [193]. It describes the modulation and channel coding technique for multi-programme DTV/HDTV services to be used for primary and secondary distribution in Fixed Satellite Service (FSS) and Broadcast Satellite Service (BSS) bands. In addition, the standard specifies DTH services for consumer Integrated Receiver Decoder (IRD), as well as collective antenna systems Satellite Master Antenna Television (SMATV) and cable television head-end stations. The major aim of the standard is to provide the specification for the adaptation of the baseband TV signals from the output of the MPEG-2 transport multiplexer to the satellite channel characteristics.

The standard is based on the Quaternary Phase Shift Keying (QPSK) modulation and concatenated forward error correction technique based on a convolutional code and a shortened Reed-Solomon (RS) code. The standard provides compatibility with Moving Pictures Experts Group-2 (MPEG-2) coded TV services [192], by synchronising a transmission structure with the MPEG-2 packet multiplex. Exploitation of the multiplex flexibility allows the use of the transmission capacity for a variety of TV service configurations, including sound and data services. All service components are Time Division Multiplexed (TDM) on a single digital carrier [193]. The functional block diagram of the system is presented in Figure 5.1.

As the DTH services via satellite are particularly affected by power limitations, the robustness against noise and interference is the main concern in DVB-S. To achieve a high power efficiency without degradation of the spectrum efficiency, the standard recommends the use of QPSK modulation concatenated

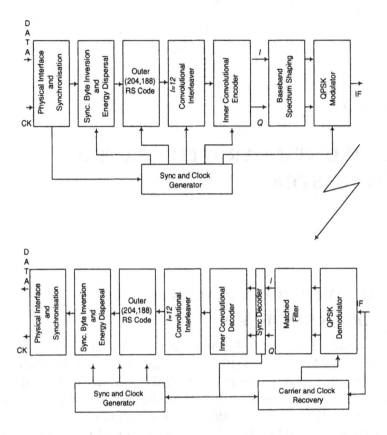

Figure 5.1.   Functional Diagram of the DVB-S System

with the convolutional and RS codes. One of the main advantages of the standard is its flexibility, which allows the optimisation of the system performance for a given satellite transponder bandwidth. The basic units on this diagram are described below:

IF interface and QPSK modulator/demodulator: these units perform the quadrature modulation and its coherent demodulation. In addition, they perform the D/A and A/D conversion, provide symbol mapping at the transmitter and "soft decision" values of I and Q components to the inner decoder. The standard recommends the use of conventional Gray-coded QPSK with absolute mapping described in Figure 5.2.

Matched filter: Prior to modulation, the I and Q components should be filtered using the filter with the square root raised cosine function defined as:

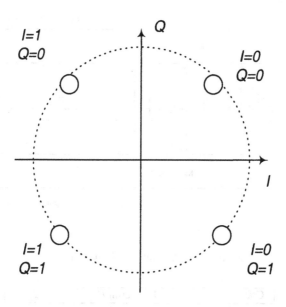

Figure 5.2.   Signal Constellation and Mapping for DVB-S

$$H(f) = \begin{cases} 1, & |f| < f_N(1-\alpha) \\ \left[\frac{1}{2} + \frac{1}{2}\sin\frac{\pi}{2f_N}\left[\frac{f_N-|f|}{\alpha}\right]\right]^{\frac{1}{2}}, & f_N(1-\alpha) \le |f| \le f_N(1+\alpha) \\ 0, & |f| > f_N(1+\alpha) \end{cases}$$

$$(5.1)$$

where $f_N = 1/2T$ is the Nyquist frequency and the roll-off factor $\alpha = 0.35$.

Carrier/clock recovery unit: this device recovers the demodulator synchronisation. The standard does not specify these units, however, it emphasises that
the probability of slips generation over the full C/N range of the demodulator
should be very low.

Inner encoder/decoder: these units perform first level error protection en-
coding/decoding. The standard specifies a range of punctured convolutional
codes, based on a rate $r = 1/2$ convolutional code with constraint length
$K = 7$, $G_1 = 171_{OCT}$ and $G_2 = 133_{OCT}$ (see Figure 5.3). The puncturing
algorithm is specified in Table 5.1.

The inner decoder should operate at an input equivalent "hard decision" BER
in the order of 10^{-2} (depending on the adopted code rate), and should produce
an output $BER = 2 \times 10^{-4}$. In addition, the inner decoder must be able to
resolve $\pi/2$ demodulation phase ambiguity.

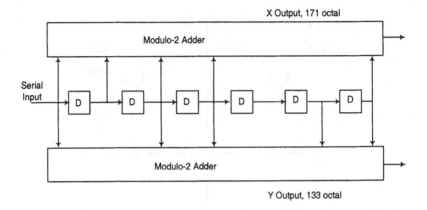

Figure 5.3.   Convolutional Encoder

No.	Code Rate	Puncturing$P$	d_{free}
1	r=1/2	I=X1;Q=Y1	10
2	r=2/3	I=X1Y2Y3; Q=Y1X3Y4	6
3	r=3/4	I=X1Y2; Q=Y1X3	5
4	r=5/6	I=X1Y2Y4; Q=Y1X3X5	4
5	r=7/8	I=X1Y2Y4Y6; Q=Y1Y3X5X7	3

Table 5.1.   Description of the Punctured Convolutional Codes

Sync byte decoder: by decoding the MPEG-2 sync bytes, this decoder provides synchronization information for the de-interleaving. It is also in a position to recover ambiguity of QPSK demodulator (not detectable by the Viterbi decoder).

Convolutional interleaver/de-interleaver: these devices allow the error bursts at the output of the inner decoder to be randomized on a byte basis in order to improve the burst error correction capability of the outer decoder. The convolutional interleaver is based on the Forney's algorithm with depth $I = 12$. The block diagram of the interleaver/de-inteleaver is shown in Figure 5.4.

Energy dispersal and energy dispersal removal: In order to comply with ITU Radio Regulations and remove long sequences of identical symbols, the pseudo random binary sequence (PRBS) generator with the polynomial:

$$p(x) = 1 + x^{14} + x^{15} \tag{5.2}$$

is used for randomisation of the MPEG-2 data and its recovery at the receiver. The block diagram of the PRBS generator is shown in Figure 5.5.

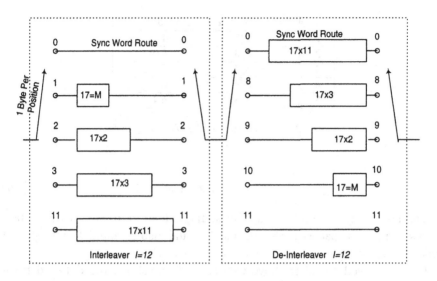

Figure 5.4.   Convolutional Interleaver/Deinterleaver

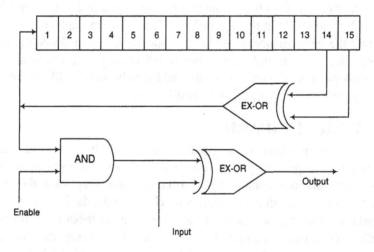

Figure 5.5.   Block Diagram of the PRBS Generator

Outer encoder/decoder: the outer encoder/decoder provide second level error protection. The (204,188) shortened RS code is derived from the original (255,239) code. The code generator polynomial is given as:

$$g(x) = (x + \lambda^0)(x + \lambda^1)(x + \lambda^2) \ldots (x + \lambda^{15}) \quad \lambda = 02_{HEX} \qquad (5.3)$$

and the field generator polynomial is given as:

No.	Inner Code Rate	Required E_b/N_0
1	$r = 1/2$	4.5
2	$r = 2/3$	5.0
3	$r = 3/4$	5.5
4	$r = 5/6$	6.0
5	$r = 7/8$	6.4

Table 5.2. Performance of the DVB-S Systems

$$p(x) = x^8 + x^4 + x^3 + x^2 + 1 \tag{5.4}$$

The code may be implemented by adding 51 all zero bytes to each randomised transport packet (188 bytes) and should also be applied to the packet synchronisation byte.

If the received signal is above C/N and C/I threshold, the Forward Error Correction (FEC) technique adopted in the standard is designed to provide a "Quasi Error Free" (QEF) quality target . The QEF means less than one uncorrected error-event per transmission hour, corresponding to Bit Error Ratio (BER) = 10^{-10} to 10^{-11} at the input of the MPEG-2 demultiplexer. Table 5.2 summarises the performance of the DVB-S standard. In this Table the values of E_b/N_0 refer to the useful bit-rate before RS coding. They also include a modem implementation margin of 0.8 dB and the noise bandwidth increase due to outer RS code $\Delta = 10 \log \frac{188}{204} = 0.36$ dB.

1.2   DSNG STANDARD

Digital television contribution applications by satellite is not intended to be received by the general public and consist of point-to-point or point-to-multipoint transmissions, connecting fixed or transportable uplink and receiving stations. The digital satellite news gathering (DSNG) standard [139] describes the modulation and channel coding system for such contribution applications by satellite. The satellite news gathering is defined as "Temporary and occasional transmission with short notice of television or sound for broadcasting purposes, using highly portable or transportable uplink earth stations..." [139]. The equipment should be capable of being set up and operated by a crew of no more than two people and limited receiving capability should be available in the uplink terminal to assist in pointing the antenna and to monitor the transmitted signal [139].

The standard assumes maximum commonality with DVB-S standard [193], by using scrambling for energy dispersal, concatenated forward error correction technique based on Reed-Solomon coding, convolutional interleaving and inner convolutional coding described in DVB-S. In general, the basic DSNG system includes (as a subset) all the transmission formats specified in [193] for QPSK modulation together with other optional high spectrum efficiency transmitter formats, based on pragmatic trellis coded 8PSK and 16QAM.

Although the use of higher order modulation techniques improves the overall efficiency of the satellite channel, their use in the DSNG systems dictates special precaution measures [139]:

1. they are more sensitive to noise and interference, therefore higher transmitted EIRPs and/or receiving antenna diameters are required;

2. they are more sensitive to linear and non-linear distortions; in particular 16QAM cannot be used on transponders driven near saturation;

3. they are more sensitive to phase noise, especially at low symbol rates; therefore high quality frequency converters should be used;

4. the possible "cycle-slips" and "phase snaps" in the receiver dictate the use of specially designed frequency conversions and demodulation carrier recovery systems.

The block diagram of the generic DSNG system is presented in Figure 5.6.

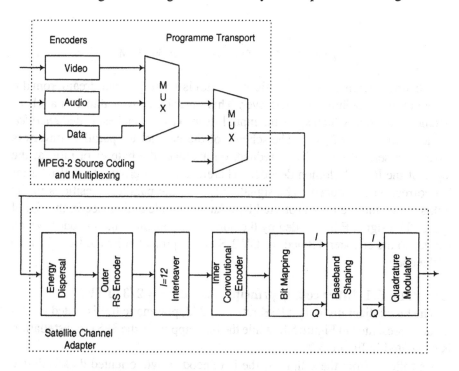

Figure 5.6. Block Diagram of the DSNG System

This generic diagram offers a number of different transmission formats, giving different trade-offs between power and spectrum efficiency. It needs to be mentioned that the use of quasi-constant envelope modulation formats (QPSK and 8PSK) allows the operation with saturated satellite power amplifiers, in single carrier per transponder configuration. However, the use of 16QAM modulation would require power back-off of the transponder.

As mentioned above, the QPSK format of the standard is similar to the DVB-S specification. In the rest of this section we will describe inner coding and modulation technique (known as pragmatic trellis coded modulation) which is specific to this particular applications.

The basic principle of the pragmatic trellis coded modulations is presented in Figure 5.7.

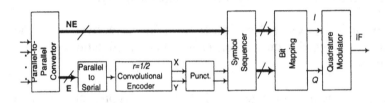

Figure 5.7.   Basic Principle of Pragmatic TCM

The input stream to the pragmatic modulator is a byte-parallel stream from the output of the convolutional interleaver. This stream is conveyed to a parallel-to-parallel converter, which splits the input bits into two branches, called *encoded (E)* and *non-encoded (NE)*. The scheme of the parallel-to- parallel converters have been selected in order to reduce, on average, the byte error-ratio at the input of the Reed-Solomon decoder. Therefore the bit error ratio (BER) after RS correction is reduced. The signals NE generate parallel transitions in the trellis code, and are only protected by a large Euclidean distance in the signal space. The signal E is encoded by the punctured convolutional encoder, which similar to the encoder defined for DVB-S and is protected by the free distance of the code.

EXAMPLE 5.1 **Inner coding principle for rate $r = 2/3$ 8PSK.**

The block diagram of the 8PSK rate $r = 2/3$, pragmatic trellis coded modulation is presented in Figure 5.8, while the bit mapping in the 8PSK constellation is presented in Figure 5.9.

As follows from these figures, the RS encoded byte oriented data is divied into the two bit oriented streams. The non-encoded (NE) stream is passed directly to the bit mapper, while the encoded (E) stream is connected to the bit

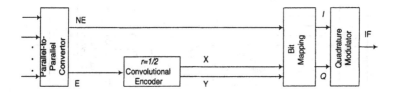

Figure 5.8.   Block Diagram of the $r = 2/3$ Pragmatic Trellis Coded 8PSK

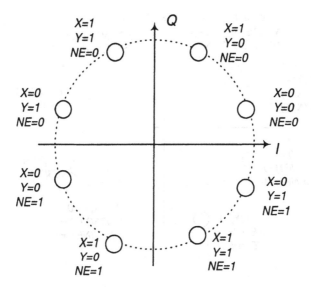

Figure 5.9.   Signal Constellation and Bit Mapping for $r = 2/3$ Pragmatic 8PSK

mapper through the $r = 1/2$ convolutional encoder. In this pragmatic trellis coded modulation scheme, 2 information bits are represented by 1 8PSK symbol providing information rate of 2 $bit/symbol$.

EXAMPLE 5.2 **Inner coding principle for rate $r = 5/6$ 8PSK**

The block diagram of this technique is presented in Figure 5.10, while the signal constellation and bit mapping is shown in Figure 5.11

In this scheme, 5 information bits are represented by 2 8PSK symbols providing information rate of 2.5 $bit/symbol$.

EXAMPLE 5.3 **Inner coding scheme for $r = 8/9$ 8PSK**.

The block diagram of this scheme is presented in Figure 5.12.

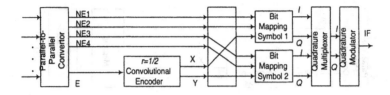

Figure 5.10.   Block Diagram of the $r = 5/6$ Pragmatic 8PSK Modulator

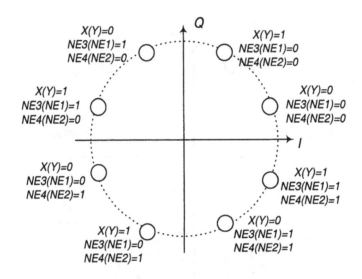

Figure 5.11.   Signal Constellation and Bit Mapping for $r = 5/6$ Pragmatic 8PSK

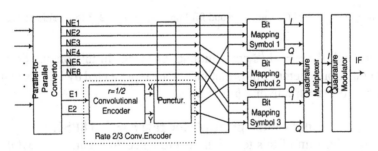

Figure 5.12.   Block Diagram of the $r = 8/9$ Pragmatic 8PSK Modulator

Signal constellation and bit mapping is similar to the previous example and is shown in Figure 5.11. As follows from these figures, this technique maps

8 binary bits into 3 8PSK symbols, providing information rate of $8/3 = 2.67$ bit/symbol.

EXAMPLE 5.4 **Inner coding principle for rate $r = 3/4$ 16QAM**

The block diagram of this technique is presented in Figure 5.13, while the signal constellation and bit mapping is shown in Figure 5.14.

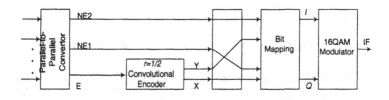

Figure 5.13. Block Diagram of the $r = 3/4$ Pragmatic 16QAM Modulator

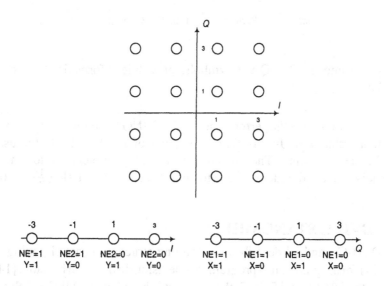

Figure 5.14. Signal Constellation and Bit Mapping for $r = 3/4$ Pragmatic 16QAM Modulator

In this scheme, 3 information bits are represented by 1 16QAM symbol, providing information rate of 3 *bit/symbol*.

EXAMPLE 5.5 Inner coding scheme for $r = 7/8$ 16QAM. Block diagram of this scheme is presented in Figure 5.15.

Signal constellation and bit mapping is similar to the previous example and is shown n Figure 5.14. As follows from these figures, this technique maps

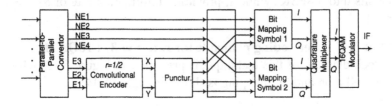

Figure 5.15.   Block Diagram of the $r = 7/8$ Pragmatic 16QAM Modulator

No.	Inner Code Rate/Modulation	Modem Implementation Margin	Required E_b/N_0
1	$r = 2/3$ 8PSK	1.0	6.9
2	$r = 5/6$ 8PSK	1.4	8.9
3	$r = 8/9$ 8PSK	1.5	9.4
4	$r = 3/4$ 16QAM	1.5	9.0
5	$r = 7/8$ 16QAM	2.1	10.7

Table 5.3.   Performance of the DVB-S Systems

7 binary bits into the 2 16QAM symbols, providing information rate of 3.5 bit/symbol.

Table 5.3 summarises the performance of the DSNG standard for higher order modulation techniques. In this Table the values of E_b/N_0 refer to the useful bit-rate before RS coding. They also include a modem implementation margin and the noise bandwidth increase due to outer RS code $\Delta = 10 \log \frac{188}{204} = 0.36$ dB.

1.3   DVB-C STANDARD

The DVB-C standard describes the framing structure, channel coding and modulation for a digital multi-programme television distribution by cable [147]. Similar to the DVB-S and DSNG, the standard is based on the MPEG-2 System Layer standard [192] with the addition of appropriate Forward Error Correction technique and quadrature amplitude modulation with 16, 32, 64, 128 or 256 constellation points.

To achieve the desired quasi error free level of error protection required for cable transmission of digital TV signals, a FEC is based on the Reed-Solomon encoding similar to the DVB-S and DSNG systems. No convolutional coding is used in cable DTV transmission systems and the protection against burst errors is achieved by the use of byte interleaving. The block diagram of the overall DVB-C system is presented in Figure 5.16.

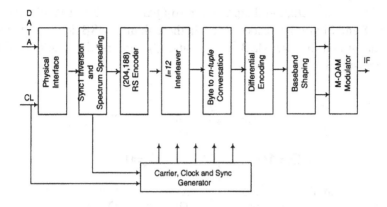

Figure 5.16.   Block Diagram of the DVB-C Modulator

Figure 5.17 describes the framing structure of the DVB-C standard. In this diagram figure a) illustrates MPEG-2 transport packet after the multiplexer, figure b) shows the randomized transport packets, figure c) presents MPEG-2 transport packet encoded with (204,188) RS code and figure d) shows inter-leaved frames of depth $I = 12$.

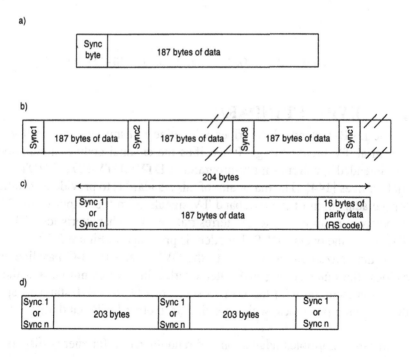

Figure 5.17.   Framing Structure of the DVB-C Signals

The DVB-C constellation diagrams for 16-QAM and 32-QAM are given in Figure 5.18 and Figure 19, respectively. In these figures MSB denotes the most significant bit and the two MSB (i.e. I_k and Q_k) define the quadrant of the transmitted signal.

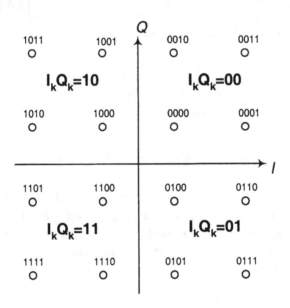

Figure 5.18.   16QAM Constellation for DVB-C Modulator

1.4   DVB-T STANDARD

The DVB-T standard describes a baseline transmission system for digital terrestrial TV broadcasting. It specifies the channel coding/modulation system intended for digital multi-programme DTV/SDTV/EDTV/HDTV terrestrial services [144]. The major aim of this standard is to provide a specification for the adaptation of the baseband TV signals from the output of the MPEG-2 transport multiplexer, to the terrestrial channel characteristics. The block diagram of the overall DVB-T modem is presented in Figure 5.20.

To maximize commonality with the DVB-S and DVB-C baseline specifications, the outer coding and outer interleaving are common, and the inner coding is common with the DVB-S standard. The DVB-T standard specifies the following processes, which are similar to other DVB standards:

1. transport multiplex adaptation and randomization for energy dispersal;

2. outer Reed-Solomon coding;

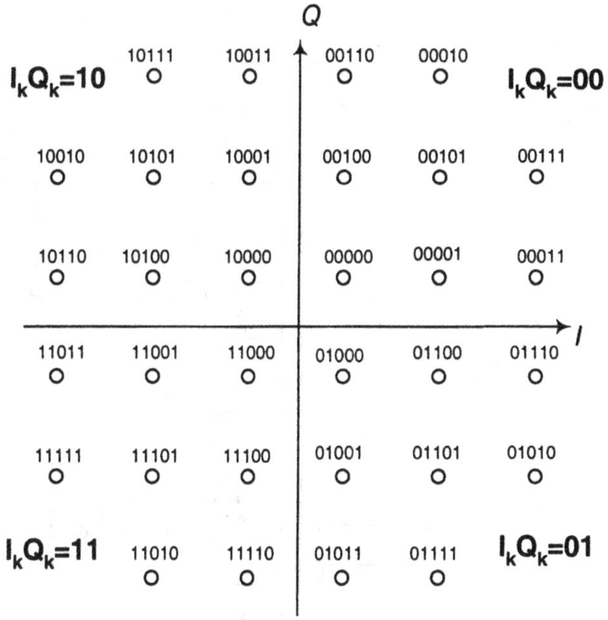

Figure 5.19. 32QAM Constellation for DVB-C Modulator

3. outer interleaving (i.e. convolutional interleaving);

4. inner coding (i.e. punctured convolutional code);

In addition, the standard specifies the following operations, which are exclusive to the DVB-T specification:

1. inner interleaving;

2. mapping and modulation;

3. Orthogonal Frequency Division Multiplexing (OFDM) transmission.

At the initial stage of DVB-T services, the system is assumed to operate within the existing VHF and UHF spectrum allocation for analogue transmissions. Therefore, the DVB-T standard is developed to provide sufficient protection against high levels of Co-Channel Interference (CCI) and Adjacent-Channel Interference (ACI) emanating from existing PAL/SECAM/NTSC services [144]. To achieve these requirements an OFDM system with concatenated error correcting coding, termed *Coded OFDM or COFDM* is specified.

Two modes of operation are defined: a "2K mode" and an "8K mode" [144]. This is referred to a number of sub-channels in the transmitted OFDM signal and the standard recommends the following:

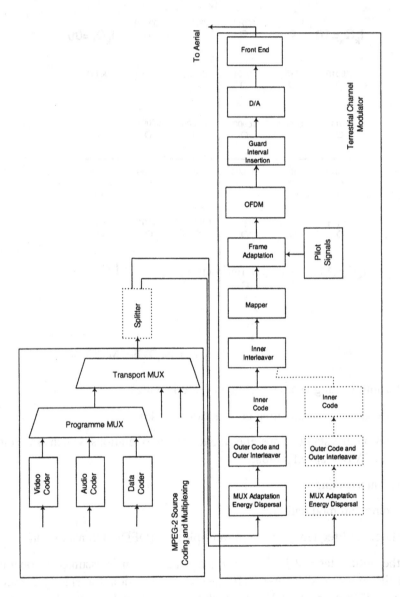

Figure 5.20.   Block Diagram of the DVB-T Modulator

- In the "2K mode", single OFDM symbol consists of $N_c = 1512$ active sub-channels. This mode is suitable for single transmitter operation and for small single frequency networks SFN networks with limited transmitter distances;

- In the "8K mode", single OFDM symbol consists of $N_c = 6048$ active subchannels. This mode can be used both for single transmitter operation and for small and large SFN networks.

The system allows different levels of QAM modulation and inner coding schemes as well as two level hierarchical channel coding and modulation. These are shown in dashed lines on the overall block diagram of the DVB-T transmitter.

The inner interleaving consists of bit-wise interleaving followed by symbol interleaving. We describe the bit-wise interleaving procedure based on the 16QAM non-hierarchical mode, as shown in Figure 5.23. In this diagram the input is demultiplexed into $v = 4$ sub-streams. In hierarchical mode, the input consists of two streams, each of which is demultiplexed into two sub-streams. Each sub-stream from the demultiplexer is processed by a separate bit interleaver. For 16QAM non-hierarchical mode, presented in Figure 5.23, there are four interleavers, labelled $I0$ to $I3$. Bit interleaving is performed only on the useful data. The block size for each interleaver is 126 bits, but the interleaving sequence is different for each mode. The block interleaving process is therefore repeated exactly twelve times per OFDM symbol of useful data in the 2K mode and forty-eight times per symbol in the 8K mode [144].

The purpose of the symbol interleaver is to map v bit words onto the 1512 (2K mode) or 6048 (8K mode) active carriers per OFDM symbol. A schematic block diagram of the algorithm used to generate the permutation function is represented in Figure 5.21 for the 2K mode and in Figure 5.22 for the 8K mode.

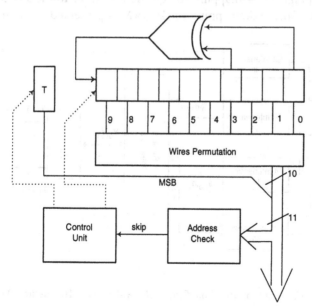

Figure 5.21.   Block Diagram of the Permutation Unit for 2k Mode

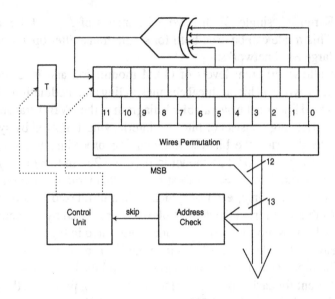

Figure 5.22.   Block Diagram of the Permutation Unit for 8k Mode

The transmitted DVB-T signal is organized in frames such that each frame consists of 68 OFDM symbols. All data carriers in one OFDM frame are either QPSK, 16-QAM, 64-QAM, non-uniform-16-QAM or non- uniform-64-QAM using Gray mapping. Gray mapping for QPSK is similar to the DVB-S and DSNG standards. Gray code mapping for 16QAM is presented in Figure 5.24.

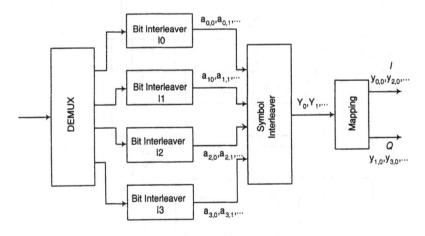

Figure 5.23.   Mapping Inpit Bits Onto Output 16QAM for Non-Hierarchical Mode

Four frames constitute one super-frame. Each OFDM symbol is composed of two parts: a useful part and a guard interval. The guard interval consists in

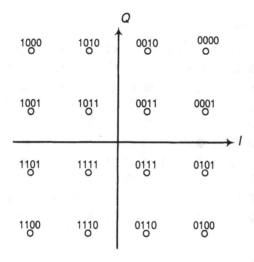

Figure 5.24. Symbol Mapping for Non-Hierarchical 16QAM

a cyclic continuation of the useful part, and is inserted before it. In addition, every OFDM frame contains pilot tones (see Figure 5.25). Some of these pilot tones are fixed in frequency and others are scattered through the symbol using a pettern of carrier positions known at the receiver. They are used for frame synchronisation, frequency synchronisation, time synchronisation, channel estimation, transmission mode identification and phase noise monitoring.

2. ADVANCED TELEVISION SYSTEM COMMITTEE [ATSC]

In Chapter 1 it was noted that Europe, the USA and Japan had developed their respective digital TV implementation plans and specifications separately but using some common technologies such as MPEG. In the USA the Advanced Television System Committee [ATSC] has been active for more than a decade and some HDTV services have been on air now since late 1998. The basic features of the RF modulation system chosen for terrestrial high definition TV broadcasting, 8 level Vestigial Side Band [8VSB], has been described in Chapter 3 so here only the specific features of the ASTC standard are outlined. The full specification is given in document A/53 of the ATSC[106].

The complete modulation process comprises the following stages, as illustrated by Figure 5.26:

1. Input Data Randomisation,

2. Reed Solomon Outer Forward Error Correction Coding,

3. Interleaving,

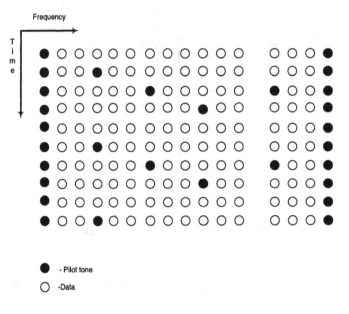

● - Pilot tone

○ -Data

Figure 5.25.   Illustration of the Frame Structure

4. Trellis Inner Forward Error Correction Coding,

5. Multiplexing of the Synchronisation Signal elements,

6. Pilot Carrier Insertion,

7. Optional Pre-Equaliser and

8. Carrier Modulation.

The data stream resulting from the Forward Error Correction [FEC] coding is formatted using two structures: A Segment Sync of 4 channel symbols every 832 symbols and a Frame Sync, comprised of a complete segment of 832 channel symbols, repeating every 313 segments, as illustrated by Figure 5.27. This structure assists the receiver in gaining reliable access to the transmitted data payload. A pilot carrier component is defined for the transmitted RF signal to assist the receiver in obtaining carrier recovery. The symbol rate is 10.76 MBaud and is related to the channel spacing of 6 MHz, commonly used in frequency planning in North America where both the VHF and UHF bands are deployed to support digital terrestrial transmissions. At three bits per symbol the data rate is in excess of 30 MBit/s but FEC and other overheads reduce the payload to near 19.3 MBit/s.

The symbol rate of 10.76.. MBaud is chosen to be 684/286 x 4.5 MHz or 684 times the NTSC line frequency. The frequency 4.5 MHz is very significant

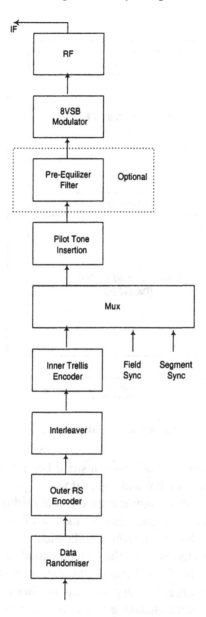

Figure 5.26.   Block Diagram of the 8VSB Transmitter

in television as explained in Chapter 1 section 1.3.1 and Chapter 2 section 2.2. This choice ensures that the spectrum of the digital 8VSB signal, even when its data content is randomised, will have some low level components at harmonics of 684/832 x 4.5/286 MHz or 12.94.. KHz which is the data segment frequency. Some of these will coincide with harmonics of the NTSC line frequency because

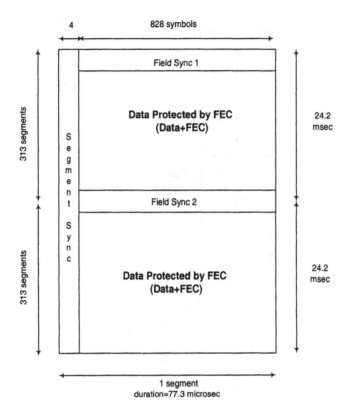

Figure 5.27.   VSB Data Frame

of their common origins and will have a small bearing upon the interference effects between the new digital and old analogue signals when they share the VHF and UHF bands. More significantly the relationship between the symbol rate and the NTSC line frequency enables the use of the interference limiting filter that is part of the 8VSB receiver specification.

The receiver is expected to adjust the incoming modulation envelope to apply correction for the effects of the channel state. A time domain equaliser [TDE], realised by means of an adaptive tapped delay line, has its tap weights adjusted using information about the channel gleaned first from the Frame and Segment Sync signals and then from the symbol decoding itself. In particular, these elements assist the time domain equaliser to adapt to the state of the channel on the long-term basis of a defined training sequence of about 66 micro-seconds duration that repeats every 24.2.. milliseconds.

The segment containing the training sequence comprises two defined pseudo-random sequences; one of length 511 symbols and one of length 63 symbols the latter being repeated three times in succession. These generate 511 + 3 x 63 = 700 symbols. There are 832 - 4 - 700 = 128 symbols remaining and these are used to carry VSB mode information [24 symbols], 104 reserved data symbols and 12 symbols that repeat data from the last 12 symbols of the previous segment in order to assist the trellis decoders. This arrangement is not adequate for the tracking of rapid changes in the channel state that might be caused by multipath and Doppler shift. Thus, additional means of driving the taps is also made available in the receiver. Decision feedback techniques, deriving information from the data decoder paths after the error correction stages, are also used to assist the equaliser to adapt to the rapid changes in channel state. This is a conditionally stable scheme because excessive rates of change in the channel state can defeat the adaptation process through aliassing caused by an inadequate rate of channel state update. Thus, the design of the equaliser must be done with great care and with due regard for the expected conditions that the channel may exhibit. The fact that the training sequence is only 700 symbols long means that it is impractical to realise a TDE that would contain the whole sequence. Typical equalisers might be 128 symbols long and so have to compute tap weights progressively as the training sequence passes through the equaliser. The computation algorithm is key to the success of this process and is not defined by the specification but left to the implementer of the receiver to choose.

The Reed Solomon Outer FEC code is constructed using MPEG packets of 188 bytes but the sync byte is suppressed thus reducing the packet length to 187 bytes. An additional 20 bytes of FEC are appended thus making a total of 207 bytes per packet. The Convolutional Interleaver, illustrated as Figure 5.28, is based on an array of 52 paths whose delays are integer multiples of 4 byte data blocks. This provides a depth of interleave of about 1/6th of a data field that is, about 4 milliseconds. This convolutional interleaver, together with randomiser, shown in Figure 5.29, is designed to give protection against expected burst errors in the channel.

The overall frame structure scheme is illustrated in Figure 3.16, while the Inner FEC is realised using a Trellis code and its encoder block diagram is illustrated as Figure 5.30 that also includes the channel symbol mapping table. The complete code is realised using twelve encoders in parallel so that symbols 0, 12, 24, 36 – are coded in one group, 1, 13, 25, 37 – in another and so on for the remaining 10 groups.

The three parallel outputs of the mapper ensure that the carrier takes 8 amplitude levels as defined by the data stream. The constellation of the carrier states is uni-dimensional.

The 8VSB standard includes a provision for the system to operate over digital cable infrastructures. The RF channel bandwidths in cable systems are generally the same as for terrestrial VHF and UHF but the conditions in the channel are more stable and so allow more C/N that in turn allows more data rate capacity. The symbol rates are double those for terrestrial channels and so the scheme

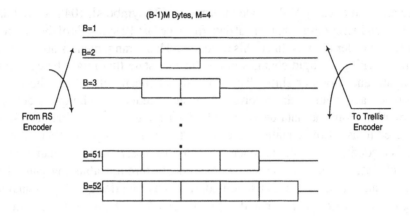

Figure 5.28.   ATSC Convolutional Interleaver

is known as 16VSB because there is twice the number of symbol states. The basic processing chain is the same as 8VSB except that the mapping is obviously based on four bits per symbol not three and the interleaver is 26 segments deep and not 52. No Trellis Inner code is employed and no interference rejection filter is specified. Both 8VSB and 16VSB are defined only for 6 MHz channels but can be scaled to others.

3.   TERRESTRIAL INTEGRATED SERVICES DIGITAL BROADCASTING [ISDB-T]

The Japanese came later to digital broadcasting than Europeans and North Americans. As a result they have had the benefit of the experiences of the former regions' standards development. This is rather like the development of colour TV where the USA decided upon NTSC in late 1953 but Europe implemented its colour systems well over ten years later having tested and explored the alternatives during the interim. Some would say that the PAL system is an improved version of the US scheme using additional features that, at the time of NTSC's choice, were not foreseen or were considered too expensive or were not implementable in the time required for service launch.

Japanese researchers chose an OFDM solution quite early in their studies but noted the features of DVB-T, standardised in late 1995 as a result of a perceived commercial need in parts of Europe, and adopted a new form that met the perceived needs of the Japanese market. Mobility was seen as a major part of the application in Japan and so the system chosen - ISDB-T - provides for this need. There are similarities with DVB-T in the standard but ISDB-T is more complex and is likely to be more expensive to realise, other things being equal, than either 8VSB or DVB-T. Nevertheless, when integrated in

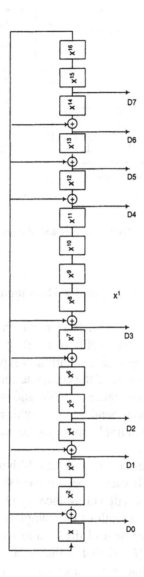

Figure 5.29.   ATSC Randomiser

silicon, this complexity will be less of a concern. The ISDB-T specification was published in September 1998[112].

The ISDB-T system uses COFDM techniques as described in Chapter 3. However, consideration of the commercial applications has led to some unique features although there is much commonality with DVB-T. For example the use

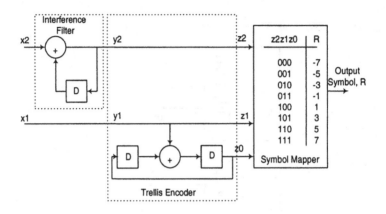

Figure 5.30.   Trellis Enocder and Mapper

of the same MPEG packet structures, outer Reed Solomon FEC, Interleaver and Inner Convolutional FEC.

Whereas the DVB-T scheme takes input data streams and spreads them across the whole occupied bandwidth of the COFDM signal, even in the Hierarchical mode, ISDB-T sub-divides the input data stream into separate streams each of which then modulates a separate part of the complete set of COFDM carriers. The scheme is known as Band Segmented OFDM and is illustrated by Figure 5.31. This allows wide or narrow band services with respectively high and low capacities but with correspondingly more ruggedness of delivery as the bandwidth and data rate reduces.

Each segment bandwidth is derived directly as 1/14th of the channel spacing so that, for example, in the 6 MHz case, it becomes $6000/14 = 428.6$ KHz. It is clear therefore that there is a band-edge allowance equivalent to one segment's bandwidth and the occupied bandwidth of the complete ensemble is approximately 5.6 MHz. The central segment of the 13 allowed may be reserved for the Partial Reception Segment [PRS] that is the most ruggedly coded and is intended for simpler receivers that only extract this segment. Where used, the PRS becomes Layer 1 of the transmission; Layers 2 and 3 are formed from groups of the remaining segments. The Layers are meant to support separate services within the same transmission. Within each segment the COFDM ensemble is self-sufficient and, because it is relatively narrow-band, it could be less susceptible than DVB-T to the selective fading of multipath events that have narrow spectral width and so are confined to one segment of the signal. DVB-T does have natural protection from such events but it is implemented differently.

The stages in the process of modulation are:

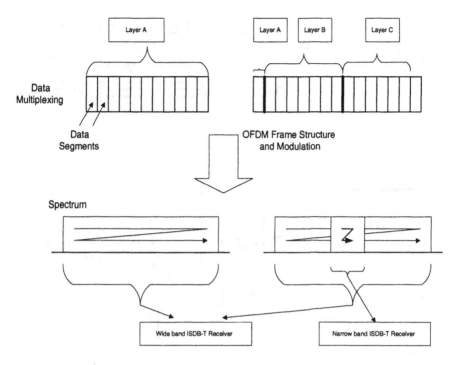

Figure 5.31. Hierarchical Transmission and Partial Reception

1. Input Data Re-multiplexing in up to three bit streams or "Layers",

2. Reed Solomon Outer Forward Error Correction Coding,

3. Randomization of data streams,

4. Interleaving,

5. Convolutional Inner Forward Error Correction Coding,

6. Mapping of layered streams to the OFDM system,

7. Time and Frequency interleaving,

8. OFDM Frame adaptation including control data insertion,

9. Inverse Fast Fourier Transform,

10. Insert Guard Interval.

These stages are illustrated in Figure 5.32.

The 13 segments can be used in various ways to convey data. Each can use a different set of modulation parameters but are grouped into one of three

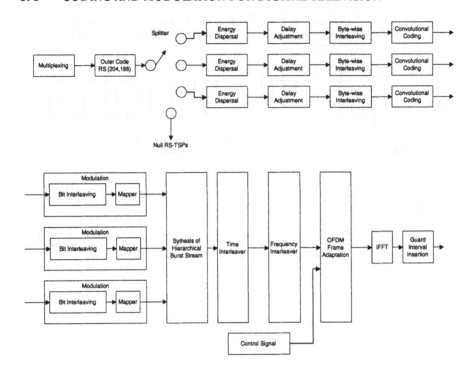

Figure 5.32.   Channel Coding and Modulation for ISDB-T

Layers accordingly. The payload capacities in each Layer can vary widely as a result and so, as a comparison, consider the 8K mode of DVB-T with FEC rate ¾, guard interval of ¼ and 64QAM that gives 22.4 MBit/s maximum payload. With ISDB-T the same FEC and modulation parameters gives a total payload of 21.9 MBit/s. These rates are both referred to an 8 MHz channel. Because the precise symbol rates and channel coding overheads are not the same the guard intervals are not identical in this example and whilst DVB-T offers 224 micro-seconds [56 micro-seconds in the 2k mode] ISDB-T offers 190 micro-seconds in mode 3 and 47 micro-seconds in mode 1. Thus DVB-T offers 2.5% more capacity and 18

As for DVB-T, frequency interleaving of data among the carriers is employed but in ISDB-T this is done only within each segment. This allows the partial reception of individual segments from the complete transmitted signal where for example ultimate ruggedness is required. If this feature is deployed the central segment of the set is used as noted above. Because the Layers may use different modulation schemes there is a delay equalisation requirement to enable the payload data to travel through the various channels available such that they arrive at the receiver in the correct order and at the correct time. Each stage of the ISDB-T process compensates for its inherent differential delays.

One significant difference between DVB-T and ISDB-T is the latter's use of time interleaving where not only are data spread dynamically across the frequency space of a segment they are also spread in time over a number of symbols within a segment. Finally, interleaving can be applied separately among grouped segments of the same type ie Layers. Each Mode of ISDB-T has a defined time interleaving format. Time interleaving requires additional memory and therefore cost for its implementation and, although it was considered for inclusion in the DVB-T standard, concerns over this cost led to its removal. It is possible to add an optional time interleaver to DVB-T for those users who have yet to choose their system. The value of the time interleaver is to provide resistance to impulsive interference where the environment is expected to be so affected.

As in DVB-T, the number of carriers that are employed can be chosen by the user. This selection is a "Mode" choice and there are three modes - 1, 2 and 3 - where the inter-carrier spacing is approximately 5.3KHz, 2.6 KHz or 1.3 KHz respectively for the 8 MHz case. Also as with DVB-T, which uses very similar values of carrier spacing [4.5 KHz for the 8k mode and 1.1 KHz for the 2k mode], this allows networks to use different symbol periods and hence different guard intervals where needed. ISDB-T is defined for 6, 7 and 8 MHz channel spacing.

The modulation formats available in IDSB-T are all those included in DVB-T ie QPSK, 16QAM and 64QAM but with the addition of DQPSK. The Layers of ISDB-T may mix modulation schemes, FEC and Interleave to provide a flexible hierarchical system.

As in DVB-T, ISDB-T uses a control and signalling channel - the Transmission and Multiplex Control Channel or TMCC - and in addition provides assistance to the carrier recovery process at the receiver by leaving some carriers of the COFDM ensemble un-modulated. These are the "pilots" whose power is boosted compared to data carriers and they can be in fixed positions or scattered in a controlled fashion over the frequency space. The TMCC is based on the use of other carriers that carry no data payload.

ISDB-T is a complex system and only an outline of its features is given above; the reader is referred to the complete specification[112] for a full appreciation of its range of capability and its implementation.

4.   NORTH AMERICAN CABLE STANDARD - DATA OVER CABLE SERVICE INTERFACE SPECIFICATIONS [DOCSIS]

The cable industry in North America is a mature and successful one. The common technical interests of the commercial operators of cable TV networks are served through CableLabs[111] that has developed a comprehensive system specification - DOCSIS - for digital cable TV networks that also allows telephony and data transmissions. Furthermore, because cable systems can readily offer return paths it is possible to support interactive services and so the specification covers means to provide this support. All the relevant protocols for the data transmission functions form the main bulk of the specification that comprises over 300 pages. The parts that relate to the RF Up- and Down-stream modulation signals are relatively short.

The cable environment is relatively stable and free of major channel defects. Noise, adjacent channel interference, intermodulation distortion and reflections are the main impairments. Simple linear modulation systems can be used to great effect and the linearity of the channels that have been designed for analogue systems generally means that there is good C/N that can be traded for digital capacity.

The cable infrastructures targeted for digital DOCSIS systems are implemented using Hybrid Fibre-Copper technologies and carry both analogue and digital signals. The characteristics of the digital modulation must therefore be benign in their effects on the sensitive analogue NTSC signals but also rugged to their converse interfering effects. A 6MHz channel plan is assumed to extend from about 50 to 864 MHz, at its fullest extent, for the down-stream direction and, for the Up-stream direction, from 5 to 42 MHz where NTSC analogue television signals may also be present. In the case of the Downstream direction the 6 MHz channel supports two slightly different symbols rates of 5.06 MBaud and 5.4 MBaud for 64QAM and 256QAM respectively. At respectively 6 and 8 bits per symbol coding efficiency the respective gross capacities are 30.4 and 43.2 MBit/s. The high order of these modulation schemes illustrates the channel linearity available. Raised cosine channel shapes are defined with 18% roll-off for 64QAM and 12% for 256QAM. Interleaving and FEC are implied but not defined except by reference to ITU-T specification J.83.

In the Up-stream direction there is more specific definition of the RF modulation system. In addition to QPSK and 16QAM in two modes using Gray and Differential symbol coding there is definition of several symbol rates - 160, 320, 640, 1280 and 2560 KBaud - and Reed Solomon FEC using byte-wise processing over GF[256] with t=1 to 10. No RS coding is also permitted as an option. Interleaving is defined and randomisation of data is required. Contention resolution for Up-stream traffic is provided by means of FDMA/TDMA systems that support a controlled framework in which the consumer's modem can offer messages to the head-end in bursted mode [TDMA] and in RF channels defined by centre frequencies [FDMA]. The modem has to be frequency agile to move from channel to channel in use.

By modifying some of the parameters to account for different cable system features, a version of the DOCSIS specification - EuroDOCSIS - has been proposed for use in Europe and this is in clear competition to the DVB-C system described above.

Chapter 6

FUTURE TRENDS IN DIGITAL TELEVISION

1. ESTIMATION OF SYSTEM PERFORMANCE

Once a standard is accepted, the manufacturers of the broadcasting equipment and consumer set-top boxes would prefer to keep the standard unchanged for a great number of years. However, for operators of DTV systems it is always essential to improve their commercial efficiency by improving the use of available bandwidth or transmitter power. Therefore, there exists a problem of finding a universal criterion for the optimisation of broadcasting systems, which should take into account both techical and commercial requirements.

The conventional criterion for communication systems optimisation is the criterion of minimum probability of error (or maximum a posterior probability) [185]. However, this criterion is applicable only for receiver optimisation, as the transmitted side is assumed to be specified. Therefore, the conventional criteria are not directly applicable to a broadcasting system, when modulation and coding schemes are to be chosen.

In [186] the following efficiency criterion has been suggested. Let our aim be to optimise a broadcasting system with information rate R bit/sec, bandwidth B Hz, and bit energy per noise E_b/N_o dB. We define the bandwidth efficiency as:

$$\gamma = \frac{R}{B} \qquad bit/sec/Hz \tag{6.1}$$

and the channel capacity as:

$$\eta = \frac{R}{C} \tag{6.2}$$

where C is the channel capacity specified as:

$$C = B \times \log_2(1 + SNR) \tag{6.3}$$

and SNR is the signal-to-noise ratio in the channel, expressed as [185]:

$$SNR = \frac{E_b}{N_0} \times \frac{R}{B} = \gamma \frac{E_b}{N_0} \tag{6.4}$$

Thus, channel capacity efficiency can be re-written as:

$$\gamma = \frac{R}{B \log_2(1 + SNR)} = \frac{\gamma}{\log_2(1 + \gamma \frac{E_b}{N_0})} \tag{6.5}$$

It follows from the Shannon theory that the maximum value for channel capacity efficiency

$$\gamma_{max} = 1 \tag{6.6}$$

Thus, for optimum communication system:

$$\gamma = \log_2(1 + \gamma \frac{E_b}{N_0}) \tag{6.7}$$

or

$$2^\gamma = 1 + \gamma \frac{E_b}{N_0} \tag{6.8}$$

and the optimum trade-off between the E_b/N_o and frequency efficiency, also known as Shannon limit, can be derived as:

$$\frac{E_b}{N_0} = \frac{2^\gamma - 1}{\gamma} \tag{6.9}$$

This is presented in Figure 6.1, which also illustrates efficiencies some of the DVB-DSNG systems as specified by ETSI [139] . It can be observed that the efficiency of the current standard is at least 4.5 dB away from the Shannon bound. Therefore, there is a theoretical potential for the improvement in the existing DVB systems. In this Section we describe different technique that could reduce this gap and provide a means for further improvements in the current standards.

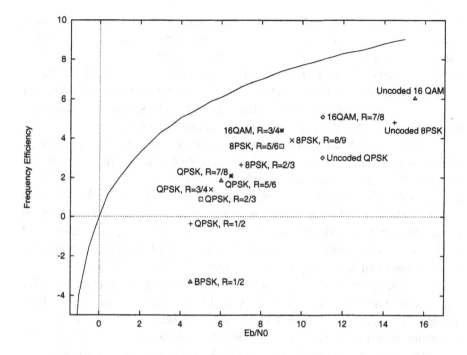

Figure 6.1.   The Shannon Limit for the current DVB-DSNG systems

It needs to be mentioned that all the techniques that will be presented in this Section are not compatible with the existing DVB standards. Therefore, their implementation in direct-to-home DTV systems may require significant efforts. However, these systems are likely to be implemented in proprietary or contribution DTV systems, where the broadcaster controls both the transmitter and receiver and cost issues are not so crucial as in the case of consumer receivers.

2.   MODULATION TECHNIQUES FOR FUTURE DTV SYSTEMS

2.1   MULTI-DIMENSIONAL MODULATION FORMATS

Multi-dimensional modulation was introduced as a means for increasing information rate without sacrificing minimum distance between the signal points or increasing signal energy [187], [188], [135]. To design such a modulation one may use either the time domain or the frequency domain or both in order to increase the number of dimensions.

The simplest explanation of such a technique can be provided for the case of 3- dimensional 8-PSK. Let us consider a sphere of radius R (same as the radius of a circle in the case of the conventional 8-PSK) with the cube embedded inside the sphere with its centre at the origin. It is apparent that the 8 points where cube touches the surface of the sphere will be maximumally separated in distance. Hence, they can be chosen as the 8 signal points for the desired 3-dimensional 8-PSK. It is clear that the minimum distance in this signal set is larger than that of the conventional 2-dimensional 8-PSK (although the signal energy remains the same) and can provide better error performance. Another example of multi-dimensional modulation is an OFDM signal in which every subcarrier could be considered as a separate dimension of the signal.

One may assume that choosing a higher dimension for the signal constellation will result in a higher energy gain. However, a signal space limited in time to an interval τ and to a one-sided bandwidth W can have no more than $N = 2\tau W$ possible dimensions [135]. For example, in QPSK the symbol duration $\tau = 2T$ (where T is the period of the incoming data stream) and the channel bandwidth can be considered to be $W = 1/T$. Therefore, the number of dimensions available within this bandwidth $N = 4$. This bound on the number of dimensions can be reached by moving from $QPSK$ to Q^2PSK.

The signal space of Q^2PSK is the same as that of twice using $QPSK$ (information rate obtainable with this type of modulation is similar to 16-PSK format). It uses two data shaping pulses and two carriers, which are pairwise quadrature in phase, to create a four-dimensional signal space and increases the transmission rate by a factor of two over QPSK. This improvement in bandwidth efficiency is similar to the improvement brought by QPSK over BPSK. However, the bit error rate performance depends on the choice of pulse pair.

The general equation for Q^2PSK modulation can be written as follows [187]:

$$S(t) = a_1(t) \cos(\frac{\pi t}{2T}) \cos(2\pi f_c t) + a_2(t) \sin(\frac{\pi t}{2T}) \cos(2\pi f_c t)$$
$$+ a_3(t) \cos(\frac{\pi t}{2T}) \sin(2\pi f_c t) + a_4(t) \sin(\frac{\pi t}{2T}) \sin(2\pi f_c t) \quad (6.10)$$

where $a_i(t)$, $i = 1, 2, 3, 4$ represent four demultiplexed streams of the input data; $R = 2/T$ is the data bit rate at the input of the modulator (it should be noticed that this bit rate is twice the bit rate of QPSK scheme), and f_c is a carrier frequency that must be a multiple of the $1/4T$:

$$f_c = \frac{n}{4T} \qquad n \geq 2 \quad (6.11)$$

Thus, at any instant, the Q^2PSK signal can be analysed as consisting of two signals; one is cosinusoidal with frequency either of $(f_c \pm \frac{1}{4T})$, the other is sinusoidal with frequency either of $(f_c \pm \frac{1}{4T})$. The separation between the two frequencies associated with either of the two signals is $1/2T$; this is the minimum spacing required for coherent orthogonality of two FSK signals of duration T. A comparison shows that the cosinusoidal part of Q^2PSK signal

exactly represents a minimum shift keying (MSK) signal [135], [191]. Therefore, the Q^2PSK signalling scheme can be thought of as consisting of two
minimum shift type signalling schemes in parallel.

It has been proved in [187] that for a wideband channel corrupted only by
AWGN all of BPSK, QPSK, and Q^2PSK belong to the same class of signalling
schemes which require 9.6 dB E_b/N_o for a bit rate of 10^{-5}. However, if the
channel is bandlimited, as it happens to be in practical broadcasting channels,
each of these schemes responds with a different level of performance degradation. For example, Q^2PSK achieves twice the bandwidth efficiency of QPSK
at the expense of 1.6 dB increase in the average bit energy (to achieve similar
rate increase with 16-PSK one would face significantly larger energy losses).
Hence, Q^2PSK is an efficient type only if the SNR is rather small, i.e. < 5dB
per dimension [189]. In the same publication two new modulation schemes,
having the same spectral density as Q^2PSK but with an increased bit rate per
symbol, have been introduced for satellite broadcasting. The first modulation
scheme (termed 8^2PSK) transmits 6 bits per symbols (64 symbols of equal
energy) by using simultaneously twice 8-PSK at two frequencies separated by
$1/(2T)$. The 64 possible waveforms are given by:

$$S(t) = \cos(\omega_c t + \frac{\pi t}{2T} + \frac{i\pi}{4}) + \cos(\omega_c t - \frac{\pi t}{2T} + \frac{j\pi}{4}) \qquad (6.12)$$

where $|t| < 2T$ and $i, j \in \{0, 1, \ldots, 7\}$. The signal space corresponding to
this modulation is the 4-dimensional signal space obtained as the direct product
of twice the 8-PSK signal space. Hence this modulation type is efficient for a
SNR up to about 10 dB per signal space dimension.

The second modulation scheme has also the same spectral density as Q^2PSK
and it transmits 5 bits per symbol. The 32 possible waveforms are given by:

$$S(t) = \cos(\omega_c t + \frac{j\pi}{4}) \times \cos(\frac{\pi t}{2T} + \frac{i\pi}{4}) \qquad (6.13)$$

where $|t| < 2T$, $i \in \{0, 1, 2, 3\}$ and $j \in \{0, 1, \ldots, 7\}$. The four-dimensional
signal space of this modulation type consists of 32 points. The minimal distance between distinct signal points is twice as large as for the first modulation
scheme (but the bit rate compared to the first modulation scheme is reduced
by a factor of $5/6$). This modulation scheme can be considered as a coded
version of the 8^2PSK modulation scheme (it also can be considered as a modulation that generates a Slepian type signal set). The efficient use of these 4-
dimensional modulation formats is associated with the specially designed error
control coding schemes operating over rings of integers.

2.2   MINIMUM SHIFT KEYING

Minimum shift keying is a modified form of offset QPSK (OQPSK) in that
I and Q channel sinusoidal pulse shaping is employed prior to multiplication
by the carrier. The transmitted MSK signal can be represented by [191]:

$$S(t) = a_n \sin(\frac{2\pi t}{4T}) \cos(2\pi f_c t) + b_n \cos(\frac{2\pi t}{4T}) \sin(2\pi f_c t) \qquad (6.14)$$

where a_n and b_n are the $n - th$ I and Q channel symbols. MSK signalling
is an example from a class of modulation techniques called continuous phase
modulation (CPM). It has a signal constellation which could be interpreted as a
time varying phasor diagram. The phasor rotates at a constant angular velocity
from one constellation point to an adjacent point over the duration of one MSK
symbol. When $a_n = b_n$ the phasor rotates clockwise and when $a_n \neq b_n$ the
phasor rotates anticlockwise.

An alternative interpretation of MSK signalling is possible in that it can be
viewed as a special case of binary frequency shift keying (BFSK) modulation
[135]. When the phasor is rotating anticlockwise the MSK symbol has a constant "high" frequency $f_H = f_c + \frac{1}{4T}$ Hz and when rotating clockwise it has a
"low" frequency $f_L = f_c - \frac{1}{4T}$ Hz. It is apparent that these frequencies, "high
frequency" and "low frequency" should be as close together in the frequency
domain as possible and still remain orthogonal over the bit time interval T. The
difference $\Delta f = f_H - f_L = \frac{1}{2T}$ is the minimum frequency that satisfies this
condition (the name of the MSK modulation is chosen due to this fact).

The MSK signals can be classified as Type-I and Type-II [132]. In Type-II
MSK the basic pulse shape is always a positive half-sinusoid . For Type-I MSK,
the pulse shape alternates between a positive and a negative half-sinusoid. For
both Type- I and Type-II MSK there is not a one-to-one relationship between
the input data and the resulting frequencies, f_H and f_L, in the MSK signal. To
get a one-to-one frequency relationship between a Type-I MSK signal and the
corresponding FSK signal, called fast frequency-shift keyed (FFSF) signal, the
data input to the Type-I MSK modulator is first differentially encoded [132].

MSK signals can be generated by using any one of several methods, as
illustrated in Figure 6.2.

Figure 6.2a shows the generation of FFSK (which is equivalent to Type-I
MSK with differential encoding of the input data). Here a simple FM-type
modulator having a peak deviation of $\Delta F = 1/(4T)$ is used.

The parallel method of generating MSK is shown in Figure 6.2b.

Figure 6.2c shows the serial method of generating MSK. In this approach,
BPSK is first generated at a carrier frequency of $f_2 = f_c - \Delta F$ and the bandpass
filtered about $f_1 = f_c + \Delta F$ to produce an MSK signal with a carrier frequency
of f_c.

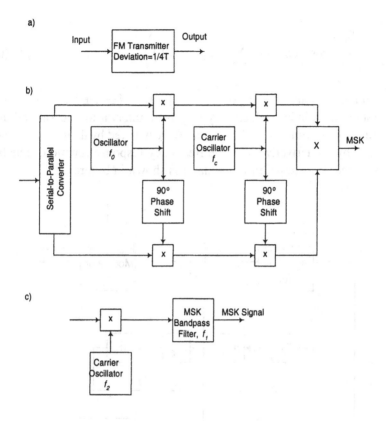

Figure 6.2.   Generation of MSK Signals

Compared to QPSK, the MSK spectrum has a broader main lobe but more rapidly decaying sidelobes, which is particularly attractive for satellite transponders with frequency division multiplexing systems. The use of MSK in these systems will allow the reducion in adjacent channel interference and improvement in the overall bandwidth efficiency. The probability of bit error for ideal MSK detection is identical to that for QPSK systems since orthogonality between I and Q channels is preserved.

2.3   CONTINUOUS PHASE MODULATION SCHEMES

The continuous phase modulation (CPM) schemes represent a class of modulation schemes where the RF signal envelope is constant and phase varies in a continuous manner. All CPM signals are described by the following equation [190]:

$$S(t) = \sqrt{\frac{2E}{T}} \cos\left(\omega_c t + 2\pi \sum_{i=0}^{n} a_i h_i q(t - iT)\right) \quad nT < t < (n+1)T \quad (6.15)$$

where data $\{a_n\}$ are $M - ary$ data symbols, $M \in \{\pm 1, \pm 3, \ldots, \pm(M-1)\}$; h_i is a modulation index, which may vary from interval to interval and $q(.)$ is the phase response function. CPM schemes are denoted by their phase response function or by its derivative $g(.)$, the frequency response function. The block diagram of the $M - ary$ CPM modulator is shown in Figure 6.3.

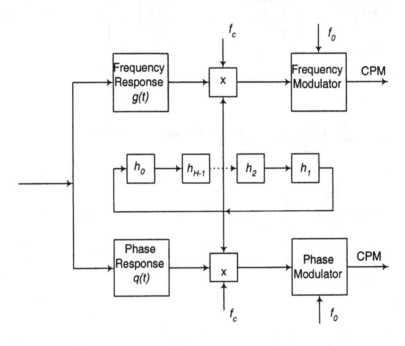

Figure 6.3.   Generation of $M - ary$ CPM Signals

The most important CPM schemes are listed below:

1. *Multi-h Modulations*: When h_i varies from interval to interval, a scheme is called $multi - h$; Otherwise, h is assumed to be fixed. Generally, a $multi - h$ scheme is one in which h_i moves cyclically through a set of indices, as shown in Figure 6.3.

2. *Full vs. Partial Response CPM*: A full response CPM modulation is one whose frequency pulse lasts one signal interval. Otherwise, the scheme is partial response.

3. *Continuous Phase FSK (CPFSK)*: CPFSK signals are continuous and signal alternatives in a symbol interval are not generally orthogonal unless h is a multiple of $1/2$. When $h = 1/2$, the CPFSK scheme is MSK.

4. *Truncated Pulse CPM*: The GMSK modulation is an example of Truncated Pulse CPM as its frequency pulses are of infinite time duration and are, therefore, time truncated in time-domain implementations.

The performance of CPM can be estimated based on the normalised Euclidean distance measure for a CPM signal [190]:

$$d^2 = \log_2 M \left[\frac{1}{T} \int_0^{NT} \left[1 - \cos \phi(t, \xi) \right] dt \right] \tag{6.16}$$

where ξ is the difference between the transmitted data sequence and the sequence used by the detector (for reference: $d_{min}^2 = 2$ for BPSK).

The upper bounds of d^2 for different values of M and h have been calculated in [190]. It has been shown that CPM has a unique feature that, for a fixed value of h, these bounds grow with M, and will actually approach ∞ as $M \to \infty$. For sufficiently large but finite SNR, the error probability can be arbitrarily small as M grows. This fact is not fundamentally new as it has been shown by Shannon that such a signal set can exist. The $M - ary$ full-response CPM scheme is one example of such a signal set. The problem is that the bandwidth of the transmitted signal also grows with M, and thus an optimum compromise between the signal bandwidth and the order of the modulation (or error performance) is required.

2.4   AN ALTERNATIVE APPROACH

In this Section we describe a technique which allows the construction of digital modulation formats with any *a priori* given number of signals. We illustrate the technique by the means of two particular examples in which 6PSK and 12QAM modulation formats have been designed. We also show that the developed techniques provide better error performance when compared with the conventional techniques.

Consider a trellis coded modulation (TCM) scheme with $M = 2^k - ary$ modulation which provides information rate of R $bit/symbol$. In such a scheme, rate $r = m/k$ error correction encoder transforms m information bits into k encoded bits, and $R = (m \log_2 M)/k = (m \log_2 2^k)/k = m$ $bit/symbol$.

Let our aim be to design an $N - ary$ ($N < M$, $N \neq 2^k$) combined coding and modulation scheme with the same information rate R bit/symbol. In order to achieve the desired goal we propose the system presented in Figure 6.4, where $kB - nN$ device represents a duobinary encoder which transfers k $B(inary)$ bits into n $N - ary$ symbols, such that

$$2^k < N^n \qquad (6.17)$$

and

$$k < n \log_2 N \qquad (6.18)$$

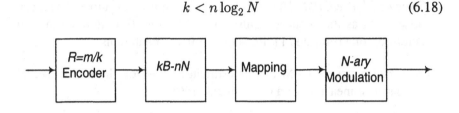

Figure 6.4.   Block Diagram of the Proposed Technique

As both k and n can be only integers, we can assume that

$$k = \lceil n \log_2 N \rceil \qquad (6.19)$$

where $\lceil . \rceil$ denotes the largest integer $\leq n \log_2 N$. For example, for $n = 3$ and $N = 9$, $k = \lceil 9.5 \rceil = 9$.

The $kB - nN$ transformation can be arranged as the natural transformation from the $k - tuple$ binary vector into the $n - tuple$ $N - ary$ number. It is apparent that the proposed scheme will provide information rate $R = m/n$ information bits per symbol and by varying m, k and n parameters, any desired rate could be achieved.

The proposed technique is illustrated by a means of the following examples.

EXAMPLE 6.1 **12-QAM with information rate** $R = 3$ $bit/symbol$

As mentioned above, the transmission of the 16QAM signals over non-linear satellite transponder requires up to 6 dB power back-off due the presence of the 4 signals with the largest energy. It is apparent that the need for power back-off could be reduced if a way of puncturing 16QAM to the 12QAM could be found. However, to the best our knowledge, the conventional techniques are associated with the loss of information due to the mapping of four binary bits into the $N = 12 - ary$ symbols. In this Section we show how the proposed algorithm can be applied in order to solve the problem.

Let our aim be to design a combined coding and 12-QAM modulation system, which could provide information rate of $R = 3$ $bit/symbol$. Following the algorithm described above we propose the technique shown in Figure 6.5.

As it follows from this Figure, the input signal is first encoded with the rate $r = 6/7$ encoder. Such an encoder can be easily derived from the conventional rate $r = 1/2$ encoder by the puncturing of the corresponding bits. At the next stage, $7 - tuple$ binary vector is transformed into the $2 - tuple$ $N = 12 - ary$ vector. Such a transformation is possible as $2^7 = 128 < 12^2 = 144$.

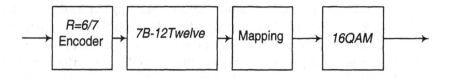

Figure 6.5.   Block Diagram of the 12QAM System

Each of the two signals from the output of the duobinary encoder are mapped into the 16QAM signal constellation in such a way that the four signal points with the largest energy (located in the corners of the constellation) will be eliminated. An example of such a constellation is shown in Figure 6.6.

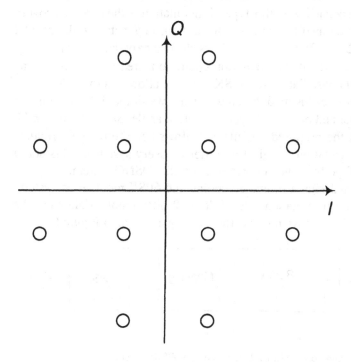

Figure 6.6.   Signal Constellation of the 12QAM Modulation

It is apparent that the proposed scheme provides information rate of $R = 6$ information bits per 2 channel symbols, i.e. 3 bit/symbol. Furthermore, the maximum signal energy in the proposed constellation is reduced by the factor of $\Delta E = 2/1.1 = 1.8$. Therefore, the proposed technique will require up to 2.5 dB less power back-off and is more suitable for the application over non-linear channels. In addition, implementation complexity of the developed system should remain almost unchanged as the conventional 16QAM modulator can be used in order to generate the desired signal set. By analysing the signal constellation of Figure 6.6 we can observe that the correct decision area for signals closest to the punctured points is increased, thus, overall error performance of the developed scheme is expected to be better than the conventional 16QAM.

The proposed scheme has been compared with the uncoded 16QAM modulation format using the developed SYNOPSYS COSSAP models. It has been found that up to 0.5 dB improvement can be obtained in AWGN channel. It is apparent that further improvement could be achieved in the non-linear channel.

EXAMPLE 6.2 **6PSK Modulation with $R = 2$ bits/symbol**

Current DVB DSNG standard specifies rate $r = 2/3$ 8PSK trellis coded modulation which allows information rate of $R = 2$ bit/symbol. One of the major problems associated with this type of modulation is the carrier recovery technique which is defined by the angle separation, $\Delta\phi$, between the nearest signals (for 8PSK $\Delta\phi = 2\pi/8 = \pi/4$). The performance of the carrier recovery systems can be improved if the separation could be increased. However, the conventional signal constellation for 8PSK does not allow this to be done.

We propose a novel 6PSK modulation which allows similar information rate of $R = 2$ bit/symbol and provides larger separation angle $\Delta\phi = 2\pi/6 = \pi/3$. We also show that the proposed modulation allows better error performance and smaller implementation margin for carrier recovery systems. This could result in the overall performance improvement of the DSNG system.

Let our aim be to design a combined coding and 6PSK modulation system, which could provide information rate of $R = 2$ bit/symbol. Following the algorithm described above we propose the technique shown in Figure 6.7.

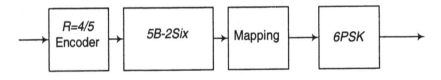

Figure 6.7.   Block Diagram of the 6PSK System

As follows from this Figure, the input signal is first encoded with the rate $r = 4/5$ encoder. Such an encoder can be easily derived from the conventional rate $r = 1/2$ encoder by the puncturing of the corresponding bits. At the next stage, a $5 - tuple$ binary vector is transformed into the $2 - tuple\ N = 6 - ary$ vector. Such a transformation is possible as $2^5 = 32 < 6^2 = 36$. The obtained signal constellation in shown in Figure 6.8.

It is apparent that the minimum Euclidean distance between the signals is increased in comparison with the conventional 8PSK signal. Therefore, the pro-

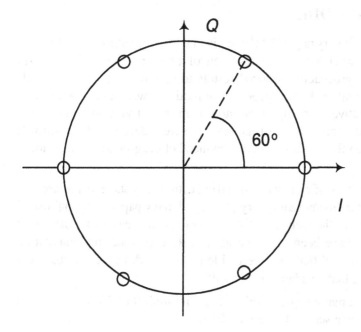

Figure 6.8. Signal Constellation of the 6PSK System

posed 6PSK could provide performance improvement, which can be estimated as:

$$\Delta E = 10 \log_{10} \frac{d_{6PSK}^2}{d_{8PSK}^2} = 10 \log_{10} \frac{1}{0.59} = 2.32 dB \qquad (6.20)$$

where d_{6PSK}^2 and d_{8PSK}^2 represent squared minimum Euclidean distances for the proposed 6PSK and conventional 8PSK signals, respectively.

It is apparent that the proposed scheme provides the information rate of $R = 4$ information bits per 2 channel symbols, i.e. 2 bit/symbol. In addition, larger phase separation will make the carrier recovery system simpler and could result in the lower implementation margin.

3.   SOFT OUTPUT DECODING TECHNIQUES

In this section we will look at some of the latest advances in coding theory that can be applied to the digital broadcasting arena. This is not meant to be a complete section and indeed at the time of writing there are new papers containing both revolutionary and evolutionary ideas constantly appearing.

We will try to give a subset that we think are ideal candidates for next generation systems given the technology and the life of the technology.

3.1   TURBO CODES

In 1993 a revolutionary paper [222] was given where the authors stated results claiming that a parallel concatenation of convolutional codes together with an iterative decoding procedure produced results to within a few tenths of a dB of the Shannon Limit [50]. This paper set in motion a wealth of research into the subject of iterative decoding, none more than those trying to replicate the results, and it was some time until the results were independently confirmed and the information theory family were convinced of the ground breaking ideas and results.

A significant amount of work was carried out at this early stage by researchers at the NASA Jet Propulsion Laboratory (JPL), and many papers were published in journals. Indeed at the web site [224] we there is an excellent repository of all the papers that have been produced at JPL. Research and implementation at JPL was so advanced that they were able to use a turbo code on board the Cassini spacecraft, that was launched in 1997.

There have been many papers published on the subject of Turbo Codes and an excellent reference set can be seen at [225].

Initially most work was being carried out on so called Turbo Convolutional Codes (TCC), those Turbo Codes that are formed by the parallel (or serial) concatenation of convolutional codes, but behind the scenes there was also substantial work being carried out on Turbo Block Codes (TBC), or what have become known as Turbo Product Codes(TPC).

Recently an excellent book on turbo codes [223] has been published that concentrates on Turbo Convolutional Codes, both parallel and serial.

We will introduce the fundamentals of both Turbo Convolutional Codes and Turbo Product Codes in the next two sections.

TURBO CONVOLUTIONAL CODES

The fundamentals of Turbo Convolutional Codes can be found in many papers [224, 222] and indeed a book [223] will provide an interesting read to the understanding of the theory of TCCs.

As stated above TCCs were invented in 1993 in the revolutionary paper [222]. This paper introduced a coding scheme based on the parallel concatenation of two simply recursive systematic convolutional codes. The parallel idea was to encode the incoming bit more than once and place at least one of the coded bits in a different part of the output by means of an interleaver. Let us consider first a recursive systematic convolutional code. In Figure 6.9 we can see a simple recursive convolutional code (more explanation of this can be found in Chapter 4).

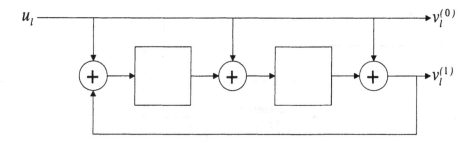

Figure 6.9.   A Simple Recursive Convolutional Code

Now consider if we construct the system as in Figure 6.10, then we see that we do not need the systematic bits to be sent across twice. We remove the systematic part of each encoder and get Figure 6.11.

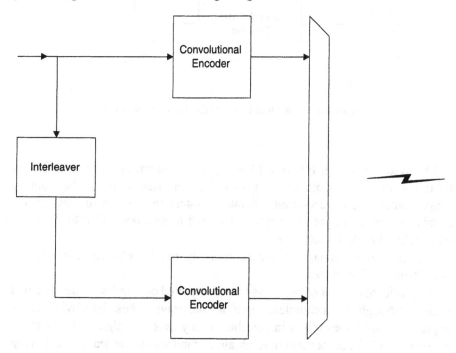

Figure 6.10.   A Simple Parallel Concatenation

Note in these we have not specified the type of interleaver and to keep it general the interleaver is there only to distribute the coded bits in different places in the transmitted stream. We will explain a little more on the types of interleaver that have been investigated later in this section.

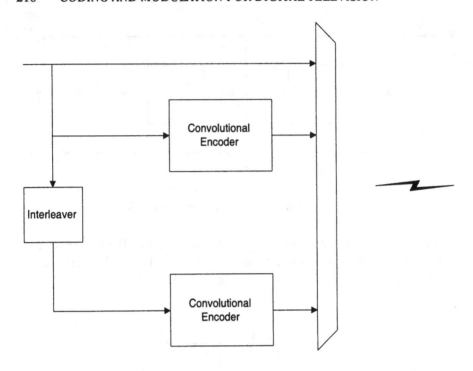

Figure 6.11.   A More Practical Parallel Concatenation

The system given in Figure 6.11 is a rate $\frac{1}{3}$ system, i.e. one information bit in and three codes bits out. This can be converted to any higher rate by simply puncturing the coded bits. A rate $\frac{1}{2}$ system can be seen in Figure 6.12. In this the coded bits are alternately punctured to give one coded bit for each information bit, thus a rate of $\frac{1}{2}$.

The interleaver that has been most commonly used is a block type interleaver (see Chapter 4 Section 9).

A significant amount of work has been completed on interleaver design. It is generally thought that a completely random interleaver is best, but obviously for implementation issues this is impossible. Many excellent algorithms exist for interleavers [235], all pertaining to obtain as random an interleaver but having an easy and implementatble construction.

Now as is normal with FEC coding schemes the power is mainly to do with the decoder and TCCs are no exception. This is the highly complex part of the system.

In high level diagrammatic form if we look at Figure 6.13 we see that we need two convolutional soft input soft output (SISO) decoders (corresponding to the two convolutional encoders in the TCC encoder). These have to soft

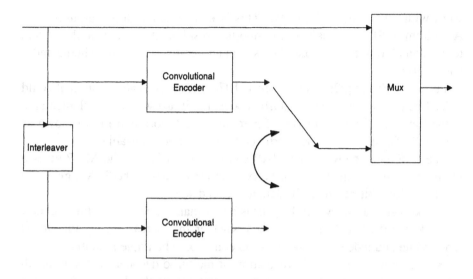

Figure 6.12. A Rate $\frac{1}{2}$ Turbo Convolutional Code

input and soft output, which is one reason why Turbo Codes at all have taken so long to be implemented.

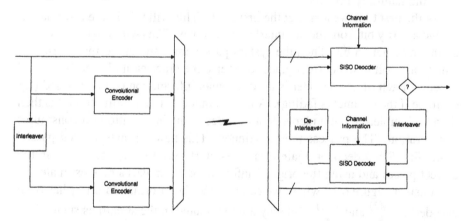

Figure 6.13. The Turbo Decoder with Turbo Encoder

As we explained in Chapter 4 Section 7, the Viterbi decoder for convolutional codes was invented in 1967 [208]. This was a breakthrough for continuous decoding of convoltuional codes which could be achieved in hardware. This decoder however only gives hard decision outputs, so for any kind of feedback decoder these are of little use. In [228] an extension of the Viterbi algorithm was introduced that would indeed give reliability estimates on the deocded word with little extra complexity. This is known as the Soft-Output Viterbi Algorithm or SOVA. This was an important result as it enabled implementers to have an algorithm that could be used in an iterative decoding scheme and it was realisable.

Further to this in [229] as far back as 1974 there was an algorithm that could indeed be used as the SISO for turbo codes, but it just was not implementable at the time. This algorithm is a Maximum A Posteriori algorithm (known as the MAP). Even now for large turbo codes this is not achievable.

The difference between the MAP and the SOVA is that the MAP gives a reliability based on the bit given every codeword whereas the SOVA gives the reliability for a bit based on the actual decoded word.

It is known that the MAP decoder is better than the SOVA for turbo codes, by up to 0.5dB, but it must be asked whether the increase in complexity is worthwhile and indeed whether it is too complex to be implementable.

We will now give a brief explanation of the turbo decoder. More in depth information can be found in one of the many tutorials that have appeared, for example [230].

If we look at Figure 6.13 then we see two SISOs each accepting input from the channel and the previous decoding attempt on the word. If we imagine that we can decode in a bit-wise fashion then the use of SOVA and the word-wise decoding naturally follows.

For the first time we arrive at the first SISO. This will take in the information bit and a parity bit from the noninterleaved encoder. This will make an estimate on the information bit. Then the system passes this estimate to the interleaver which places it in the correct place to receive the other parity bit at the second decoder. Each time a decoder is used, channel information can be utilised (for example if the channel is fading). A new estimate of the information bit is then found. Completing two independent decodes of the information bit constitutes one iteration. The process then continues. This new estimate is then passed to the first SISO decoder again (via a de-interleaver to make sure it is in the correct place) and using the original information bit makes a new estimate.

Consider Figure 6.14 where we can see that the information bit I_1 has been encoded to $C_1^{(1)}$ and $C_1^{(2)}$, and they are distributed in the stream as such.

Figure 6.14.   The Interleaved Stream

Then the turbo docoder will work as in Table 6.1. As can be seen the information bit (I_1) and the first coded bit ($C_1^{(1)}$) go into the first SISO. A new estimate of the information bit $(\hat{I}_1)$ is found. This new estimate together with the second coded bit $(C_1^{(2)})$ then goes into the second SISO, and a further estimate is found $(\hat{\hat{I}}_1)$.

$$
\begin{aligned}
I_1 C_1^{(1)} &\rightarrow \hat{I}_1 \\
\hat{I}_1 C_1^{(2)} &\rightarrow \hat{\hat{I}}_1
\end{aligned}
$$

Table 6.1. The decoding for one iteration of a turbo code

This iterative process can be run as many times as required. Obviously in hardware to have this value varying would be difficult due to synchronisation and buffer sizes so generally it is expected that a compromise would be made for the number of iterations compared to the complexity compared to the expected performance.

TURBO BLOCK CODES

The foundations of turbo block codes can be traced back to [226] when the ideas of product codes were being developed. For this book we will denote the turbo block codes as turbo product codes (TPC). Indeed in the book [203] there are ideas of the iterative decoding nature being introduced.

The main paper to introduce TPCs in the light of the development of TCCs was [227]. In this it was shown that TPCs could perform under certain conditions as well as if not better than TCCs.

We will now give a brief introduction to the workings of TPCs. First of all consisder two dimensional TPCs. We will consider the use of systematic block codes for each of the rows and columns. Note that all the rows are encoded with one particular encoder and all the columns are encoded with a potentially different encoder. In their two dimensional form the TPCs can be thought of as a matrix where we would encode the rows and the columns to form a larger matrix. This can be seen in Figure 6.15. Here the k_x information bits from each row are encoded to form n_x coded bits on each row, and then the k_y bits from each column are encoded to give n_y coded bits for the rows.

The rate of this type of TPC is

$$
R = \frac{k_x k_y}{n_x n_y} \tag{6.21}
$$

as the information block is of size $k_x k_y$ and the total size of the coded bits is $n_x n_y$. It is known that if we have the row encoders as (n_x, k_x, d_x) and the column encoders as (n_y, k_y, d_y) then the overall TPC will be $(n_x n_y, k_x k_y, d_x d_y)$. It is this value of $d_x d_y$ that provides an insight into the power of TPCs.

For simple (but powerful) codes the construction can be based upon simple block codes. In fact if we consider Extended Hamming Codes then it is known

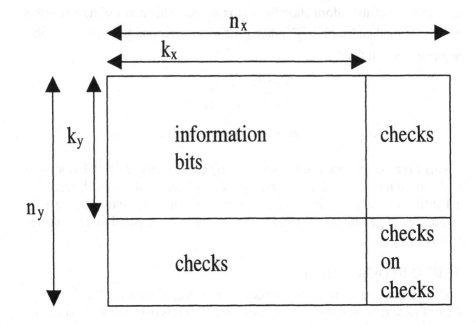

Figure 6.15.   Two Dimensional TPC

that these can be encoded and (hard decision) decoded very easily. Consider if we have a $(16, 11, 4)$ Extended Hamming Codes for the row and column codes, then we could end up with a $(256, 121, 16)$ TPC. This gives a rate $\frac{121}{256} = 0.473$ code. At first sight this might appear to place quite harsh restrictions on the rate but if we simply insert zeros to arrange the rate as required then as the constituent codes are systematic these zeros can be removed before sending and re-inserted before decoding thus enabling us to obtain almost any rate.

We will now demonstrate the encoding procedure with a specific example. Consider that we have a two dimensional $(64, 16, 16)$ TPC formed from the $(8, 4, 4)$ Extended Hamming Codes. We have 16 data bits as in Table 6.16. The way we would normally arrange these is in the obvious way of just filling up the matrix from the input stream one row at a time.

$$D_{0,0} \quad D_{0,1} \quad D_{0,2} \quad D_{0,3}$$
$$D_{1,0} \quad D_{1,1} \quad D_{1,2} \quad D_{1,3}$$
$$D_{2,0} \quad D_{2,1} \quad D_{2,2} \quad D_{2,3}$$

Figure 6.16.   The information bits for $(64, 16, 16)$ TPC

We will first encode all the rows to form Table 6.17.

$D_{0,0}$	$D_{0,1}$	$D_{0,2}$	$D_{0,3}$	$E_{0,4}$	$E_{0,5}$	$E_{0,6}$	$E_{0,7}$
$D_{1,0}$	$D_{1,1}$	$D_{1,2}$	$D_{1,3}$	$E_{1,4}$	$E_{1,5}$	$E_{1,6}$	$E_{1,7}$
$D_{2,0}$	$D_{2,1}$	$D_{2,2}$	$D_{2,3}$	$E_{2,4}$	$E_{2,5}$	$E_{2,6}$	$E_{2,7}$
$D_{3,0}$	$D_{3,1}$	$D_{3,2}$	$D_{3,3}$	$E_{3,4}$	$E_{3,5}$	$E_{3,6}$	$E_{3,7}$

Figure 6.17. Intermediate stage for encoding $(64, 16, 16)$ TPC

Next we encode all the columns (including the parity check columns just formed) to obtain Table 6.18.

$D_{0,0}$	$D_{0,1}$	$D_{0,2}$	$D_{0,3}$	$E_{0,4}$	$E_{0,5}$	$E_{0,6}$	$E_{0,7}$
$D_{1,0}$	$D_{1,1}$	$D_{1,2}$	$D_{1,3}$	$E_{1,4}$	$E_{1,5}$	$E_{1,6}$	$E_{1,7}$
$D_{2,0}$	$D_{2,1}$	$D_{2,2}$	$D_{2,3}$	$E_{2,4}$	$E_{2,5}$	$E_{2,6}$	$E_{2,7}$
$D_{3,0}$	$D_{3,1}$	$D_{3,2}$	$D_{3,3}$	$E_{3,4}$	$E_{3,5}$	$E_{3,6}$	$E_{3,7}$
$E_{4,0}$	$E_{4,1}$	$E_{4,2}$	$E_{4,3}$	$E_{4,4}$	$E_{4,5}$	$E_{4,6}$	$E_{4,7}$
$E_{5,0}$	$E_{5,1}$	$E_{5,2}$	$E_{5,3}$	$E_{5,4}$	$E_{5,5}$	$E_{5,6}$	$E_{5,7}$
$E_{6,0}$	$E_{6,1}$	$E_{6,2}$	$E_{6,3}$	$E_{6,4}$	$E_{6,5}$	$E_{6,6}$	$E_{6,7}$
$E_{7,0}$	$E_{7,1}$	$E_{7,2}$	$E_{7,3}$	$E_{7,4}$	$E_{7,5}$	$E_{7,6}$	$E_{7,7}$

Figure 6.18. Final stage for encoding $(64, 16, 16)$ TPC

So we have here, for example, the first column encoder has information bits

$$[D_{0,0}D_{1,0}D_{2,0}D_{3,0}] \tag{6.22}$$

and this is encoded to the codeword

$$[D_{0,0}D_{1,0}D_{2,0}D_{3,0}E_{4,0}E_{5,0}E_{6,0}E_{7,0}] \tag{6.23}$$

using an $(8, 4, 4)$ Extended Hamming Code.

Then all the bits of the whole matrix are modulated and transmitted across the channel. The ordering for transmission is chosen by the user, but obviously the main point is to be able to reassemble all the bits at the decoder into the same place as when they were encoded.

Now if we turn our interest to the decoder we get to the familiar position of needing a soft input soft output (SISO) decoder. Of course considering our example the standard Extended Hamming Code decoder will not suffice and we need to consider SISO block decoders.

As a first instant there are possibilities stemming from the paper by Chase [231], giving what is now known as the Chase Algorithm, which involves multiple decoding of the perturbed codewords from the recieved codeword and then using information from all decoding to give some reliability decision on the chosen one. Variations of this algorithm have been investigated since the paper was published.

TPC are easy to encode and decode and they can be implemented at a low cost for very high data rates. One such implementation is shown in Figure 6.19, which illustrates the OC-3 TPC, capable of operating at up to 250 Mbits/second.

Figure 6.19.   The AHA OC-3 TPC ASIC

Another idea would be to use a trellis system to obtain soft decisions. Considering the ideas of [233] and [232] it is possible for small block codes to produce a full trellis decoding and for larger block codes a sub trellis that will give a sub optimal but still useful decoding algorithm.

Using any of these that are implementable we can then decode the TPCs. We will explain briefly the general idea of iterative decoding of block codes.

Consider again the (64, 16, 16) TPC. We have 64 bits of coded data entering the decoder. We need to first arrange this in the correct way before we can decode it. We build up the two dimensional array once again. So we would get a picture like Table 6.18, but each element would be a quantised value from the demodulator. We then decode all 8 x direction codewords to give a new estimate of the received values giving a new matrix. Next we use the new matrix and decode all 8 y direction codewords to give a second estimate. This decoding first on the x direction and then on the y direction denotes one iteration. The process can then continue for as many iterations as required.

Of course when considering the implementation of this one needs to consider how to parallelise this process because we could have 8 separate x direction decoders and 8 separate y direction decoders, but we need to have all the information available to do this so some kind of pipeline/parallel system might be achievable.

3.2 OTHER ITERATIVE DECODING

Of course the sections above on Turbo type codes are completely incompatible with the existing standards and so their use would involve a next generation or rewrite to a new standard. Therefore well before Turbo codes were known about, researchers had to look at the current standards and try to improve them without breaking the standard. For coding and modulation the standards only specify the encoding/transmitting side. There was therefore room for improvement on the receiver/decoder side. Of course for applications like DTV this would involve new set top boxes, but the old set top boxes would still work, if the enhancement was made backward compatible. These systems at the moment are still in an exploratory stage and mainly on the benches of the researchers as the gains have not been so significant to warrant completely new designs of set top boxes.

SOFT INPUT SOFT OUTPUT CONVOLUTIONAL CODES

As mentioned in the section on Turbo Convolutional codes, there exist now SISO for convolutional codes. The two main contenders we have met are the BCJR or MAP algorithm and the SOVA algorithm. However the complexity of these has been prohibitive until recently for use on the size of convolutional codes that are used in DTV standard. There have been implementations for SOVA together with a soft input Reed-Solomon decoder to give about 0.5dB improvement to the DVB-S standard as seen in [234]. Further to this a feedback scheme was employed to allow the Reed-Solomon code to pass knowledge of decoding (uncorrectable errors) back to the Viterbi algorithm so that a new decoding using the Viterbi algorithm could be performed. This can be seen in Figure 6.20.

Of course the challenge with the system is that it stops becoming continuos and so if we are to re-decode using this extra information we will need to buffer the incoming stream. Also there is a problem of how many times do we need to re-decode? This type of system can be thought of as an approximation to full serial concatenation with iterative decoding, but it must be noted that the full iterative procedure is not employed in this scheme.

This once again has obtained small gains, but the increase in performance has not warranted its inclusion in decoder systems due to the complexity increase.

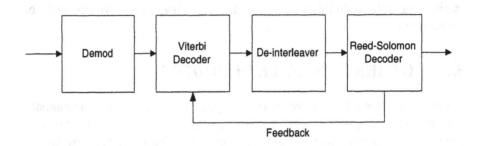

Figure 6.20.   A system employing RS feedback

SOFT INPUT SOFT OUTPUT BLOCK CODES

We have seen in the section on Turbo Block Codes (or as we called them Turbo Product Codes or TPC), that there do indeed exist implementable SISO block decoders. There have been implementations of SISO block decoders but these have been for small block length. To consider a SISO for the DVB RS code with parameters $(204, 188, 8)$ then we have a (binary) block length of $204 \times 8 = 1632$, or working over F_{256} a length 204. This is a large code and to build a trellis would involve a lot of work. A full description of trellis decoding can be found in [233], and so we will not explain it.

We will however give a brief explanation of the Chase Algorithm [231] to exemplify the multiple decodings and therefore the problems associated with implementation.

In Figure 6.21 we can see a very high level flowchart of the Chase Decoder. Here we can see that the Log Likelihood Values (LLR) are taken from the demodulator and the channel and passed to the decoder. We are only considering the use of LLRs to ease the mathematics of the system, the direct probabilities would work equally well. The least reliable bits of the codeword are then perturbed to give a number of test codewords, and then multiple decodings of these take place. Then using the original LLR values and the decoding we obtain codeword metrics and further we can obtain bit metrics from the A Posteriori block. These are then passed out or either subtracted from the original to obtain the extrinsic LLR which would be used in a further iteration.

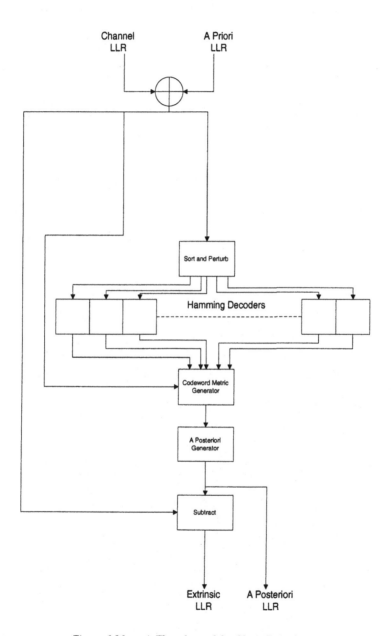

Figure 6.21. A Flowchart of the Chase Decoder

References

[1] Zworykin, A.K., "The Iconoscope – A modern version of the electric eye", *Proc. IRE, Vol 22, No 1* , pp 16-32, January 1934.

[2] Abramson, A., "Pioneers of Television - Philo Taylor Farnsworth", *SMPTE Journal*, pp 770-784, November 1992.

[3] Engstrom, E.W., "A study of TV image characteristics", *Proc. IRE*, Vol 21, No 12, pp 1631-1651, December 1932.

[4] Wengstrom, W.H., "Notes on TV definition", *Proc. IRE*, Vol 21, No 9, pp 1317- 1327, September 1933.

[5] Engstrom, E.W., "Determination of frame frequency for television in terms of flicker characteristics", *Proc. IRE*, Vol 24, No 4, pp 295-310, April 1935.

[6] Kell, R.A., Bedford, A.V., and Trainer, M.A., "An experimental television system", *Proc. IRE*, Vol 22, No 11, pp 1246-1265, November 1934.

[7] Kell, R.A., Bedford, A.V., and Trainer, M.A., "Scanning sequence and repetition rate of television images", *Proc. IRE*, Vol 24, No 4, pp 559-575, April 1936.

[8] Kell, R.D., Bedford, A.V., and Fredendall, G.L., "A Determination of optimum number of lines in a television system", *RCA Review*, Vol 5, No1, pp 8-30, 1940.

[9] Mertz, P. and Gray, F.A., "A Theory of Scanning and its relation to the characteristics of the transmitted signal in Telephotography and Television", *Bell System Technical Journal*, Vol 13, No3, pp 464-515, 1934.

[10] Bedford, A.V., "A Figure of merit for television performance", *RCA Review*, Vol 3, No 7, July 1938.

[11] Wheeler, H.A. and Loughren, A.V., "The Fine Structure of Television Images", *Proc. IRE*, Vol 26, No 5, May 1938.

[12] Wilson, J.C., "Channel Width and Resolving Power in Television Systems", *Journal of the Television Society*, Vol 2 No 2 Part II, pp 397-420, June 1938.

[13] Burns, R.W., *A History of Television Broadcasting*, IEE History of Technology Series, Number 7, Peter Peregrinus, London, ISBN 0 86341 079 0, 1986.

[14] Briggs, A., *The History of Broadcasting in the UK*, Volume 2, "The Golden Age of Wireless", Chapter 5, Oxford University Press, 1965.

[15] Burns, R.W., *A Biography of A D Blumlein*, IEE, ISBN 0 852 96773 X, 1999.

[16] Alexander, R.C., *The life and works of A D Blumlein*, Focal Press, ISBN 0 240 51577 3, 1999.

[17] Baker, W.J., *A History of the Marconi Company*, London, Menthuen, 1970.

[18] Field, H.M., *History of the Atlantic Telegraph*, London, Sampson Low, 1866.

[19] Bright, C., *Submarine Telegraphs*, London, Crosby Lockwood, 1898.

[20] Standage, T., *The Victorian Internet*, London, Phoenix, ISBN 0 75380 703 3, 1999.

[21] McDermot, E.T. and Clinker, C.R., *History of the Great Western Railway*, Ian Allan, London , Volume 1 1833-63, Chapter 12, [1927 rev. 1964].

[22] Platt, A., *The Life and Times of Daniel Gooch*, Allan Sutton, Gloucester, England, ISBN 0 86299 317 2, 1987.

[23] Nyquist, H., "Certain Topics in Telegraph Transmission Theory", *Trans. AIEE*, Vol 47, No 6, pp 617-644, April, 1928.

[24] McFarlane, M.D., "Digital Pictures 50 Years ago", *Proc. IEEE*, Vol 60, No 7, pp 768 – 770, July 1972.

[25] Mason, A.G., Drury, G.M., and Lodge, N.K., "Digital Television to the Home - When will it Come?", *IEE Conference Publication* 327, IBC, pp 51 - 57, 1990.

[26] "The White Book", UK PAL Specification, Revised Edition, Published jointly by DTI/IBA/BBC/BREMA, 1984.

[27] Townsend, B., *PAL Colour Television*, Cambridge University Press, Second Edition, 1985.

[28] Baldwin, J.L.E., Stalley, A.D., Coffey, J.A., Greenfield, R.L., Lever, I.R., and Taylor, J.H., "DICE: The first Intercontinental Digital Standards Converter", *RTS Journal*, September/October 1974.

[29] Baldwin, J.L.E., Stalley, A.D. and Kitchen, H.D., "A Standards Converter using Digital Techniques", *RTS Journal*, January/February, pp 3-11, 1972.

[30] Lucas, K. and Windram, M.D., "Standards for Broadcasting Satellite Services", *IBA Technical Review, Standards for Satellite Broadcasting*, Number 18, ISSN 0308-423 X, pp 12- 27, March 1982.

[31] Malcher, A.T., "A TDM System for ENG Links", *IEE Conference Publication 286*, International Broadcasting Convention [IBC], IBC, pp 271-274, 1984.

[32] Dalton, C.J., and Malcher, A.T., "Communications between Analogue Component Production Centres", *SMPTE Journal*, Volume 97, pp 606-612, 1988.

[33] Chippindale, P. and Franks, S., "DISHED – The Rise and fall of British Satellite Broadcasting", Simon and Schuster, London, ISBN 0 671 71077 X, 1991.

[34] Griffiths, E., and Windram, M.D., "Wide Screen Broadcast TV – A Commercial Reality", *IEE Conference Publication 327*, International Broadcasting Convention [IBC], IBC, pp 137-144, 1990.

[35] EBU Specification Tech 3258, October 1986.

[36] ETSI Specification for the D2-MAC System, ETS 300 250.

[37] ETSI Specification for the D-MAC System: ETS 300 355.

[38] ETSI Specification for the HD-MAC System: ETS 300 350.

[39] Annegarn, M. et al, *Philips Technical Review*, Volume 43, Number 8, pp 197-212, August 1987.

[40] Ninomiya, Y., et al, *IEEE Transactions on Broadcasting*, Volume BC-33, Number 4, pp 130-160, December 1987.

[41] Oliphant, A., "An Extended PAL System for High Quality Television", *BBC Research Department Report BBC RD* 1981/11, December 1981.

[42] ETSI Specification for the PALPlus System: ETS 300 731.

[43] Carbonara, C., "HDTV: An Historical Perspective – 1884-1940", *HDTV World Review*, Vol 1 No 1, Winter, 1990

[44] Carbonara, C., "HDTV: An Historical Perspective – 1940 to the Present", *HDTV World Review*, Vol 1 No 2, Spring, 1991

[45] Carbonara, C, "The Current history of HDTV", *HDTV World Review*, Vol 1 No 3, Winter, 1991

[46] Drury, G.M., "HDTV - How many Lines?", *BKSTS Image Technology Journal*, pp 16-20, January 1990.

[47] Sandbank, C.P., Editor, *Digital Television*, Wiley, ISBN 0 471 92360 5, 1990.

[48] Nyquist, H., "Certain Factors affecting Telegraph Speed", *BSTJ*, Vol 2, p 324, April 1924.

[49] Hartley, R.V.L., "Transmission of Information", *BSTJ*, Vol 6, p 535, July 1928.

[50] Shannon, Claude E., "A Mathematical Theory of Communication", *BSTJ*, Vol 26, pp 379-423, July 1948 and October 1948 pp 623-656 See also Shannon, C.E. and Weaver, W., "The Mathematical Theory of Communication", University of Illinois Press, ISBN 0 252 72548 4, 1949.

[51] Reeves, A.H., British Patent No 535,860, 1938.

[52] Oliver, B.M., Pierce, J.R. and Shannon, C.E., "The Philosophy of PCM", *Proc. IEEE*, pp 1324-1331 , November 1948.

[53] Kretzmer, E.R., "Statistics of Television Signals", *BSTJ*, Vol 31, pp 751-763, July 1952.

[54] Goodall, B.M., "Television by Pulse Code Modulation", *BSTJ*, Vol 30, pp33-49, January 1951.

[55] Cutler, C.C., "Differential Quantisation of Communication Signals", US Patent No 2,605,361, Application Date June 29th 1950. The Patent for Differential PCM.

[56] Harrison, W.C., "Experiments with Linear Prediction in Television", *BSTJ*, Vol 31, pp 764-783, July 1952.

[57] Elias, P., "Predictive Coding", *Trans. IRE on Information Theory*, Vol IT-1, pp16-33, March 1955.

[58] Gouriet, G.G., "Bandwidth Compression of a Television Signal", *Proc. IRE*, Vol 104, Part 8, pp257-272, May 1957.

[59] Pratt, W.K., "A Bibliography on Television Bandwidth Reduction Studies", *IEEE Trans on Information Theory*, Vol IT-15, pp 114, January 1967.

[60] Edson, J.O. and Henning, H.H., "Broadband Codecs for experimental 224Mbit/s PCM Terminal", *BSTJ*, VOL 44 No 9, pp 1887-1941.

[61] Smith, B.D., "An unusual electronic Analogue to Digital Conversion method", *Trans. IRE*, Vol PGI-5, pp 156-161, June 1956.

[62] Waldhauer, F.D., "Folding Converter" US Patents 3187325 and 3161868 and British Patent 1040614

[63] "Pulse Code Modulation for High Quality Sound Signal Distribution: Coding and Decoding", *BBC Research Department Report*, El-18, 1968/29

[64] Dorward, R.M., "Aspect of the Quantisation Noise associated with the Digital Coding of Colour Television Signals", *Electronics Letters*, Volume 6, Number 1, pp 5-7, January 8th, 1970.

[65] "Pulse Code Modulation of video signals: Subjective study of coding parameters", *BBC Research Department Report*, El-57, BBC RD 1971/40, October 1971.

[66] "Pulse Code Modulation of video signals: Subjective tests on acceptable limits for timing jitter in decoded analogue samples", *BBC Research Department Report*, EL-58, 1971/42, November 1971.

[67] Devereux, V.G. and Wilkinson, G.C., "Digital Video: Effect of PAL decoder alignment on the acceptable limits of timing jitter", *BBC Research Department Report*, EL-74, BBC RD 1974/1, February 1974.

[68] Moore, T.A., "Digital Video: Number of bits per sample required for reference coding of luminance and colour difference signals", *BBC Research Department Report*, BBC RD 1974/42, December 1974.

[69] Devereux, V.G., "Digital Video: Sub Nyquist sampling of PAL colour signals", *BBC Research Department Report*, BBC RD 1975/4, January 1975.

[70] Weston, M., "A PAL/YUV digital system for 625 line international connections", *BBC Research Department Report*, BBC RD 1976/24, September 1976.

[71] Devereux, V.G., "Permissible timing jitter in broadcast PAL colour television signals", *BBC Research Department Report*, BBC RD 1977/14, March 1977.

[72] Stott, J.H., and Phillips, G.J., "Digital Video: Multiple Sub-Nyquist Coding", *BBC Research Department Report*, BBC RD 1977/21, June 1977.

[73] Slavin, K.R., "The subjective effects of random bit errors in YUV component video signals", *BBC Research Department Report*, BBC RD 1985/17, December 1985.

[74] Drury, G.M., "Digital Transmission of Television Signals", *EBU Review-Technical*, Number 225, Geneva, pp 3-15, October 1987.

[75] ETSI, "Network Aspects; Digital Coding of component television signals for contribution quality applications in the range 34 - 45 MBit/s", ETS 300 174, November 1992. Also published as: *CCIR Recommendation 723*, "Transmission of component coded digital television signals for contribution quality applications at the third hierarchical level of CCITT Recommendation G.702", CCIR Documents, XVIIth Plenary, Dusseldorf, 1990, Volume XII, pp 54 - 67. Now adopted by ITU-T as recommendation J.81.

[76] CCIR Recommendation 721, "Transmission of component coded digital television signals for contribution quality applications at bit rates near 140 MBit/s", *CCIR Documents*, XVIIth Plenary, Dusseldorf, Volume XII, pp 68 - 79, 1990. Now adopted by ITU-T as Recommendation J.80.

[77] ITU-T Recommendation G 702

[78] ITU-T Recommendation G 703

[79] DirecTv web address: www.directv.com

[80] DVB web address; www.dvb.org

[81] ATSC web address: www.atsc.org

[82] MPEG web address: www.mpeg.org

[83] ISO/IEC International Standard, IS 10918-1, JPEG

[84] Reimers, Prof. U., "The European Project on Digital Video Broadcasting - Achievements and Current Status", *IEE Conference Publication* 397, IBC 1994, Amsterdam, pp 550 - 556, September 16-20th 1994.

[85] Forrest, J.R., Ed., *Electronics and Communication Journal*, IEE, Volume 9 Number 1 , Special Issue on DVB, February 1997.

[86] Comminetti, M. and Morello, A., "Direct-to-Home Digital Multi-Programme Television by Satellite", *IEE Conference Publication* 397, IBC 1994, Amsterdam, pp 358 - 365, September 16- 20th 1994.

[87] ETSI, "Digital broadcast systems for television sound and data services - Framing structure, Channel coding and modulation for 11/12 GHz satellite services", ESTI Specification ETS 300 421, November 1994.

[88] ETSI, "Digital broadcast systems for television sound and data services - CATV systems", ETSI Specification ETS 300 429, November 1994.

[89] ETSI, "Digital broadcast systems for television sound and data services - Framing structure, Channel coding and modulation for SMATV", ETSI Specification ETS 300 473, December 1994.

[90] ETSI, "Digital broadcast systems for television sound and data services - Specification for conveying ITU-R System B Teletext in Digital Video Broadcasting [DVB] Systems", ETSI Specification ETS 300 472, December 1994.

[91] ETSI, "Digital broadcast systems for television sound and data services - MMDS systems using frequencies above 10 GHz", ETSI Specification ETS 300 748, January 1996.

[92] ETSI, "Digital broadcast systems for television sound and data services - MMDS systems using frequencies below 10 GHz", ETSI Specification ETS 300 749, January 1996.

[93] ETSI, "Digital broadcast systems for television sound and data services - Framing structure, Channel coding and modulation for Digital Terrestrial Television", ETSI Specification ETS 300 744, February 1996.

[94] ETSI, "Digital broadcast systems for television sound and data services - Specification for Service Information [SI] in Digital

Video broadcasting [DVB] Systems", ETSI Specification ETS 300 468, February 1996.

[95] ETSI, Report ETR 211, "DVB Guidelines on the use of MPEG Programme Specific Information [PSI] and DVB Service Information [SI]"

[96] ETSI, Report ETR 154, DVB Guidelines on the use of MPEG

[97] ISO/IEC International Standard, IS 13818-x, MPEG-2

[98] ETSI Report ETR 289, "Support for Scrambling and Conditional Access [CA] within Digital Broadcasting Systems", October 1996.

[99] CEN/CENELEC, "Standard for Common Receiver Interface", EN 50221, April 1996.

[100] Common Scrambling/De-Scrambling Algorithm, Distribution Agreement, DVB Document A011, October 1995.

[101] "Guidelines for Implementation and use of the Common Interface for DVB decoder applications", Document A025, May 1997.

[102] Technical Specification of DVB Simulcrypt, Document A028, May 1997.

[103] DVB Commercial Module MHP Sub-Group Document, MHP 045, Rev 12, September 1998.

[104] DVB Technical Module Document, TM 2208 Rev 3, February 2000.

[105] ATSC document A/52, "Digital Audio Compression [AC3] Standard", 1995.

[106] ATSC document A/53, "ATSC Digital Television standard", 1995.

[107] ATSC document A/54, "A Guide to the Use of the ATSC Digital Television standard", 1995.

[108] ATSC document A/58, "Harmonisation with DVB SI in the use of the ATSC Digital Television Standard", 1996.

[109] ATSC document A/65, "Program and System Information Protocol for Terrestrial Broadcast and Cable", 1997

[110] ATSC document A/70, "Conditional Access System for Terrestrial Broadcast", 1999.

[111] www.cablelabs.com

[112] www.dibeg.org

[113] Bylanski, P. and Ingram, D. G. W., *Digital Transmission Systems*, Second Edition, Peter Peregrinus, ISBN 0 906048 42 7, 1980.

[114] SMPTE Specifications may be found at www.smpte.org

[115] SMPTE Specification 244M "TV: System M/NTSC Composite Video Signals – Bit Parallel Interface", 1995.

[116] SMPTE Specification 259M, "TV - 10 bit 4:2:2 Component and 4Fsc Composite Digital Signals – Serial Digital Interface", 1997.

[117] Oliphant, A., Taylor, K.J. and Misson, N.T., "The visibility of noise in System I Colour Television", *BBC Research Department Report*, BBC RD 1988/12, November 1988.

[118] Drewery, J.O., "The Filtering of Luminance and Chrominance signals to avoid cross colour in a PAL colour system", *BBC Engineering*, pp 8-39, September 1976.

[119] ISO/IEC International Standard, IS 11172-x, MPEG-1

[120] Benoit, H., *Digital Television – MPEG1, MPEG2 and Principles of the DVB System*, Arnold, ISBN 0 340 69190 5 and 0 471 25810 4, 1997.

[121] ISO/IEC, International Standard, IS14496, MPEG-4.

[122] Proposed SMPTE Standard 319M, "Transporting MPEG-2 Recoding Information through 4:2:2 Component Digital Interfaces", February 2000.

[123] Proposed SMPTE Standard 327M, "MPEG-2 Video Recoding Data Set", February 2000.

[124] Proposed SMPTE Standard 329M, "MPEG-2 Video Recoding Set – Compressed Stream Format", February 2000.

[125] ETSI Specification ETS 300 800, "Interaction Channel for Cable TV Distribution Systems [CATV]", July 1998.

[126] ETSI Specification ETS 301 195, "Interaction Channel through GSM", February 1999.

[127] ETSI Specification ETS 301 193, "Interaction Channel through DECT", July 1998.

[128] ETSI Specification ETS 301 801, "Interaction Channel through PSDN/ISDN", July 1997.

[129] ETSI Report TR 101 201, "Interaction Channel for SMATV Distribution Systems: Guidelines for versions based on satellite and co-axial sections", October 1997.

[130] ETSI Specification ETS 301 199, "Interaction Channel for Local Multipoint Distribution System [LMDS] Distribution Systems", June 1999.

[131] Proakis J. *Digital Communications*, McGraw-Hill International, 1995.

[132] Leon W. Couch II. *Digital and Analog Communication Systems*, Macmillan Publishing Company , New York, 1993.

[133] Steele R., Hanzo L. *Mobile Radio Communications*, John Whiley & Sons Ltd. 1999.

[134] Glover I.A., Grant P.M. *Digital Communications*, Prentice Hall, 1998.

[135] Webb W., Hanzo L. *Modern Quadrature Amplitude Modulation*, IEE Press, New York, 1995.

[136] Ungerboeck G. "Channel coding with multilevel/phase signals", *IEEE Transactions on Information Theory*, vol.28, No.1, pp.55-67, 1982.

[137] Ungerboeck G. "Trellis coded modulation with redundant signal sets, Part I and II", *IEEE Communications Magazine*, vol.25, No.2, pp.5-21, 1987.

[138] Honary B., Markarian G. "New optimisation technique of error propagation over M-PSK modem". *Electronics Letters*, vol.26, No.6, pp.408-410, 1990.

[139] EN 301 210. Digital Video Broadcasting (DVB); Framing structure, channel coding and modulation for Digital Satellite News Gathering (DSNG) and other contribution applications by satellite, ETSI, December 1998.

[140] Gitlits M., Lev A. *Theoretical Backgrounds of Telecommunications*, Moscow, Radio & Sviaz, 1985, (in Russian)

[141] Smith J.G. "Odd-bit quadrature shift keying", *IEEE Transactions on Communications*, vol.COM-23, No.3, pp.385-389, 1975.

[142] Markarian G., Huggett A., Mason A. "Novel high-order modulation techniques for future DVB-S systems", *IBC'99*, Amsterdam, 1999.

[143] Miyauchi K., Seki S., Ishio H. "New Techniques for generating and detecting multilevel signal formats", *IEEE Transactions on Communications*, vol.COM-24, No.2, pp.263-267, 1976.

[144] EN 300 744: "Digital Video Broadcasting (DVB); Framing structure, channel coding and modulation for digital terrestrial television", ETSI Specification.

[145] European Patent No.0506400A2/Mitsuaki Oshima. Signal Transmission System, 1992.

[146] Markarian G., Haigh P., Harding W. "Improvements in or Relating to Hiarachical Coding. TandbergTV Patent Disclosure No.99-22, 1999.

[147] EN 300 429. Digital Video Broadcasting (DVB); Framing structure, channel coding and modulation for cable systems, ETSI Specification, April 1998.

[148] "VSB Transmission System Grand Alliance: Technical Details". Monograph Published by the Grand Alliance, USA, December 1994.

[149] "Guide to the use of the ATSC Digital Transmission Standard", by Advanced Television System Committee, USA, October 4, 1995.

[150] Gita R., Sgrignoli G. "ATSC Transmission System: VSB Tutorial", Preprint, Zenith, USA, 4.1.97.

[151] Sgrignoli G. "Measuring Peak/Average Power Ratio of the Zenith/AT&T DSC-HDTV Signal with Vector Signal Analyzer". *IEEE Transactions on Broadcasting*, Vol.39, No.2, pp.255-264, 1993.

[152] Sgringnoli G.,Eilers C. "Digital Television Transmission Parameters - Analysis and Discussion", *IEEE Transactions on Broadcasting*, Vol.45, No.4, pp.365-385, December 1999.

[153] Chang R.W. "Synthesis of band Limited Orthogonal Signals for Multichannel Data Transmission", *Bell Systems Technical Journal*, Vol. 45 , pp.1775-1796, December 1966.

[154] Salzberg B.R. "Performance of an efficient parallel data transmission system", *IEEE Transactions on Communications*, Vol.15, pp.805-813, December 1967.

[155] Mosier R.R., Clabaugh R.G. "Kineplex, a Bandwidth Efficient Binary Transmission System", *AIEE Transactions*, Vol.76, pp.723-728, January 1958.

[156] US Patent No. 3 4884555. "Orthogonal Frequency Division Multiplexing", filed 14 November 1966, Issued 6 January 1970.

[157] Weinstein S.B., Ebert P.M. "Data Transmission by Frequency Division Multiplexing Using the Discrete Fourier Transform", *IEEE Transactions on Communications*, Vol.19, pp.628-634, October 1971.

[158] Zou W.Y., Wu Y. "COFDM: an overview", *IEEE Transactions on Broadcasting*, Vol.41, No.1, pp.1-8, March 1995.

[159] Sari H., Karma G., Jeanclaude I. "Transmission techniques for Digital Terrestrial TV Broadcasting", *IEEE Communications Magazine*, Vol.33, pp.100-109, February 1995.

[160] van Nee R., Prasad R. *OFDM for Wireless Multimedia Communications*, ARTECH House Publishers, Boston, 2000.

[161] Porter G.C. "Error Distribution and Diversity performance of a Frequency Differential PSK HF modem", *IEEE Transactions on Communications*, Vol.16, pp.567-575, August 1968.

[162] Zimmerman M.S., Kirsch A. "The AN/GSCC-10 (KATHRYN) Variable Rate Data Modem for HF Radio", *IEEE Transactions on Communications*, Vol.15, pp.197-205, April 1967.

[163] Radio Broadcast Systems; Digital Audio Broadcasting (DAB) to Mobile, Portable and Fixed receivers. ETS 300 401, ETSI, Valbonne, France, March 1996.

[164] Edfors O., Sandell M., van de Beek J-J., Landstrom D., Sjoberg F. "An Introduction to Orthogonal Frequency Division Multiplexing", *Research report*, Lulea University of technology, September 1996.

[165] Zyablov V.V., Korobkov D.L, Portnoy S.L. "Highspeed information transmission in real channels", Moscow, Radio and Svyaz, 1991, 288 p. (*in Russian*)

[166] Zyablov V.V., Korobkov D.L, Portnoy S.L."Concatenated signal-to-code constructions in Gaussian channel with intersymbol interference", *Trudi NIIR*, No 1, p.5-12, , 1984 (*in Russian*)

[167] Bingham J.A.C. "Multicarrier Modulation for Data Transmission: An Idea Whose Time Has Come". *IEEE Communications Magazine*, Vol.28, pp.5-14, May 1990.

[168] Peled A., Ruiz A. "Frequency Domain Data Transmission Using Reduced Computational Complexity Algorithms". *Proceedings of the IEEE International Conference on Acoust., Speech, Signal Proceeding*, Denver CO., pp.964-967, 1980.

[169] Blahut R. *Fast Algorithms for Digital Signal Processing*, Reading MA., Addison-Wesley, 1985.

[170] Vahlin A., Holte N. "Optimum Finite Duration Pulses for OFDM". *IEEE Transactions on Communications*, Vol.44, No.1, pp.10-14, January 1996.

[171] Le Floch B., Avard M., Berrou C. "Coded Orthogonal Frequency -Division Multiplexing". *Proceedings of IEEE*, Vol.83, No.6, pp.982-996, June 1995.

[172] Schreiber W.F. "Advanced Television Systems for Terrestrial Broadcasting: Some Problems and Some Solutions", *Proceedings of the IEEE*, Vol.83, No.6, pp.958-981, June 1995.

[173] Pollet T.M., van Bladel M., Moeneelaey M. "BER Sensitivity of OFDM Systems to Carrier Frequency Offset and Wiener Phase Noise", *IEEE Transactions on Communications*, Vol.43, No.2/3/4, pp.191-193, 1995.

[174] Pauli M., Kuchenbecker H.P. "Minimisation of the Intermodulation Distortion of a Nonlinearly Amplified OFDM Signal". *Wireless Personal Communications*, Vol.4, No.1, pp.93-101, January 1997.

[175] Gardner F. "Demodulator reference recovery techniques suited for digital implementation. *Final Report ESTEC Contract No.6847/86/NL/DG*, ESA, August 1988.

[176] Banket V.L., Melnik A.M.. "Sistemy vosstanovlenia nesuchei pri kogerentnom prieme discretnyh signalov (Carrier recovery techniques for coherent reception of digital signals)". *Zarubejnaia Radioelectronika*, No.2, pp.28-49, 1983 (*in Russian*).

[177] Stifler J.J. *Theory of Synchronous Detection*. Englewood Cliffs, NJ: Prentice-Hall, p.247, 1971.

[178] Simon M.K. "Optimum receiver structures for phase-multiplexed modulations". *IEEE Transactions on Communications*, Vol.26, No.6, pp.865-872, 1978.

[179] Franks L.E. "Carrier and bit synchronisation in data communication - A tutorial review". *IEEE Transactions on Communications*, Vol.28, No.8, pp.1107-1121, 1980.

[180] Leclert A., Vandamme P. "Universal Carrier Recovery Loop for WASk and PSK Signal Sets", *IEEE Transactions on Communications*, Vol.31, No.1, pp.130-136, January 1983.

[181] Markarian G. "Recovery of a Carrier Signal from a Modulated Input Signal". GB Patent Application No.GB9804182.5, 2 March 1998.

[182] Feher K. *Wireless Digital Communications: Modulation and Spread Spectrum Applications*, Prentice Hall,1995.

[183] Kingsbury N.G. "Transmit and receive filters for QPSKsignals to optimise the performance of linear and bandlimited channels". *IEE Proc.Pt.F.*, Vol.133, No.4, pp.345-355, July 1986.

[184] DVB-RCS 002(Rev.5). Digital Video Broadcasting (DVB) interaction channel for Satellite Distribution Systems, ETSI, April 2000.

[185] Wozencraft J., Jacobs I. *Principles of Communication Engineering*, Wiley, NY, 1965.

[186] Zuko A., Falko A., Panfilov I., Banket V., Ivanchenko P. *Noise Immunity and Efficiency of Communication Systems*, Moscow, Radio and Sviaz, 1985, (*in Russian*).

[187] Saha D., Birdsall T. "Quadrature-Quadrature Phase Shift Keying", *IEEE Transactions on Communications*, COM-37, No.5, pp.437-448, May 1989.

[188] Markarian G., *et all.* "Multi-dimensional Modulation Technique", USSR patent No.1830623, 1993.

[189] "Technical Assistance for CDMA Communication System Analysis", ETH, *ESTEC Contract No.8696/89/NL/US*, Bimonthly Progress Report, 31 August 1991.

[190] Anderson J., AUlin T., Sundberg C.E. *Digital Phase Modulation*, Plenum Press, NY, 1986.

[191] Pasupathy S. "Minimum Shift Keying: A Spectral Efficient Modulation", *IEEE Communications Magazine*, No.8, pp.14-22, 1979.

[192] ISO/IEC 13818-1. "Coding of Moving Pictures and Associated Audio". June 1994.

[193] EN 300 421. "Digital Video Broadcasting (DVB): Framing Structure, Channel Coding and Modulation for 11/12 GHz Satellite Services". ETSI, August 1997.

[194] Ramsey J.L. "Realisation of optimal interleavers", *IEEE Trans Info. Theory, Vol. IT-16 No. 3*, pp. 338-345, May 1970.

[195] Forney G.D. "Burst-correcting codes for the classic burst channel", *IEEE Trans Commun. Technol., Vol. COM-19*, pp. 772-781, October 1971.

[196] R.W. Hamming. "Error Detecting and error correcting codes", *Bell Sys. Tech. J.*, 29, pp. 147-160, 1950.

[197] W.W Peterson. "Encoding and error correction procedures for the Bose-Chaudhuri codes, *IRE Trans Inform. Theory, Vol. IT-6*, pp. 459-470, September 1960.

[198] Hocquenghem, A. "Codes correcteurs d'erreurs", *Chiffres (Paris), Vol. 2*, pp. 147-156, Spetember 1959.

[199] Bose, R.C. and Ray-Chauduhuri, D.K., "On a class of error correcting binary group codes", *Information and Control, Vol. 3*, pp. 68-79, March 1960.

[200] Shannon Not yet

[201] Gorenstein, D. and Zierler, N. "A class of cyclic error-correcting code in p^m symbols", *J. Soc. Ind. Appl. Math.*, pp. 107-214, June 1961.

[202] Reed, I.S. and Solomon, G. "Polynomial codes over certain finaite fields", *J. Soc. Ind. Appl. Math., Vol. 8*, pp. 300-304, June 1960.

[203] Lin S. and Costello, Jr. D. J., *Error Control Coding: Fundamentals and Applications*, Prentice Hall, Inc., Englewood Cliffs, N.J. 07632. 1983.

[204] Lee, L.H.C., *Convolutional Coding - Fundamentals and Applications*, Artech House, Inc. 685 CantonStreet, Norwood, MA 02062, 1997.

[205] Lee, L.H.C., *Error Control Block codes for communication engineers*, Artech House, Inc., Norwood, MA 02062. 2000

[206] Lidl, R. & Niederreiter, H., *Finite Fields, Encyclopedia of Math. and its Applications, 20*, Addison-Wesley; Reading, Mass. 1983

[207] Vanstone, S.A. & Ooorschot, P.C., *An Introduction to Error Correcting Codes with Applications*, Kluwer Academic Publishers, 1989.

[208] Viterbi, A.J., "Error Bounds for Convolutional Codes and an Asymptotically Optimum Decoding Algorithm", *IEEE Trans Information Theory, Vol. IT-13, No2*, pp. 260-269, April 1967.

[209] Cain, J.B., Clark, G.C. and Geist, J.M., "Punctured Convolutional Codes of Rate $(n-1)/n$ and Simplified Maximum Likelihood Decoding", *IEEE Trans Information Theory, Vol. IT-25, No1*, pp. 97-100, Jan 1979.

[210] Lee, P.J., "Construction of Rate $(n-1)/n$ Punctured Convolutional Codes with Required SNR Criterion", *IEEE Trans Information Theory, Vol. IT-36, No9*, pp. 1171-1174, Sept 1988.

[211] Begin, G. and Haccoun, D., "High Rate Punctured Convolutoional Codes Structure Properties and Construction Techniques", *IEEE Trans Information Theory, Vol. IT-37, No12*, pp. 1813-1825, Dec 1989.

[212] MacWilliams, F.J. and Sloane, N.J.A., *The Theory of Error-Correcting Codes*, North-Holland Mathematical Library, Elservier Science B.V. 1977.

[213] Peterson, W.W. and Weldon, Jr., E.J., *Error Correcting Codes*, The MIT Press, Cambridge, Massachusetts, 1961.

[214] Berlekamp, E.R., "On Decoding Biary Bose-Chaudhuri-Hocquenghem Codes", *IEEE Trans Inf Theory, Vol. IT-11*, pp. 577-580, October 1965.

[215] Berlekamp, E.R., *Algebraic Coding Theory*, McGraw-Hill, New York, 1968.

[216] Massey, J.L., "Shift Register Synthesis and BCH Deconding", *IEEE Trans Inf Theory, Vol. IT-15, No. 1*, pp. 196-198, January 1972.

[217] Chien, R., "Cyclic Decoding Procedures for Bose-Chaudhuri-Hocquenghem Codes", *IEEE Trans Inf Theory, Vol. IT-10*, 1964.

[218] Ungerboeck, G., "Channel coding with multilevel/phase signals", *IEEE Trans Inf Theory, Vol. IT-28*, pp. 55-67, 1982.

[219] Ungerboeck, G., "Trellis-coded modulation with redundancy signal sets", Part 1 and 2, *IEEE Communications Magazine, Vol. 25, No. 2*, pp. 5-21, 1987.

[220] Viterbi, A.J., Wolf, J.K., Zehavi, E. and Padovani, R., "A Pragmatic Approach to Trellis-Coded Modulation", *IEEE Communications Magazine, Vol. 27, No. 7*, 1989.

[221] ETSI, "Digital Video Broadcasting (DVB); Framing Structure, channel coding and modulation for Digital Satellite News Gathering (DSNG) and other contribution applications by satellite", ETSI Specification ETS 301 210, July 1998.

[222] Berrou, C., Glavieux, A. and Thitimajshima, P., "Near Shannon Limit Error-Correcting Coding and Decoding: Turbo Codes", *Proceedings of the 1993 International Conference on Communications*, pp. 1064-1070, 1993.

[223] Heegard, C. and Wicker, S.B., *Turbo Coding*, Kluwer Academic Publishers, 101 Philip Drive, Assinippi Park, Norwell, Massachusetts, 0261. 1999.

[224] http://tmo.jpl.nasa.gov/progress_report/

[225] http://www.csee.wvu.edu/ mvalenti/tc-bibliography.html

[226] Elais, P., "Erro-Free Coding", *IRE Trans Info Theory, PGIT-4*, pp. 29-37, September 1954.

[227] Pyndiah, R., Glavieux, A., Picart, A. and Jacq, S., "Near Optimum Decoding of Product Codes", *Proc IEEE GLOBECOM*, pp. 339-343, 1994.

[228] Hagenauer, J. and Papke, L. "A Viterbi Algorithm with Soft-Decision Outputs and its Applications", *Proc IEEE Globecom*, pp. 47.1. - 47.1.7, 1989.

[229] Bahl, L.R., Cocke, J., Jeinnek, F. and Raviv, J., "Optimal Decoding of Linear Codes for minimizing symbol error rate", *IEEE Trans Information Theory, Vol. IT-20*, pp. 248-287, 1974.

[230] Hagenauer, J., "The Turbo Priciples: Tutorial Introduction and State of the Art", *Proceedings of the International Symposium on Turbo Codes and Related Topics*, Brest, France, 1997.

[231] Chase, D., "A CLass of Algorithms for Decoding Block Codes with Channel Measurement Inforomation", *IEEE Trans Info Theory, Vol. IT-18, No. 1*, pp. 170-182, 1972.

[232] McEliece, R.J., "On the BCJR trellis for linear block codes", *IEEE Trans INfo Theory, Vol. 42, No. 4*, pp. 1072-1092, 1996.

[233] Markarian, G. and Honary, B., *Trellis Decoding of Block Codes - A Practical Approach*, Kluwer Academic Publishers, 101 Philip Drive, Assinippi Park, Norwell, Massachusetts, 0261. 1997.

[234] Wicker,S.B. and Bhargava, V.K., *Reed-Solomon Codes and their Application*, IEEE Press, The Institute of Electrical and Electronic Engineers, Inc., New York, 1994.

[235] Berrou, C., Douillard, C. and Jezequel, M., *Designing Turbo Codes for Low Error Rates*, IEE Colloquium on Turbo Codes in Digital Broadcasting - Could it double capacity, London, 22 November 1999.

Index